The CNC Insider:

Fundamentals of the Computer Numerical Control Machine Industry

By Raul Lepez

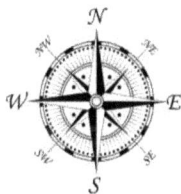

Table of Contents

Chapter 1

Basics and Benefits of CNC Machinery

Basics of CNC Machinery

CNC stands for Computer Numerical Control. CNC machines are basically conventional machines such as mills and lathes controlled by components and computer programs that consist of alphanumeric data. Their various functions, input parameters, and motions are automatically carried out based upon the information that has been programmed into its central machine control that contains the computer and executes the inputted commands. The commands which are inputted then cause mechanisms to control main functions and parts of the equipment such as turning a spindle on or off to increasing or decreasing the speed or adjusting the depth of a particular cut that must be made. CNC machinery allows for virtually any command to be placed into the program, making it ideal for manufacturing processes of all varieties.

When purchasing a CNC machine or training to operate these potentially complex machines, it's vital to have a basic understanding of the components which comprise most CNC machines. A basic CNC machine consists of several key components: the programmed parameters of the part to be made, the machine control unit (MCU), and the machine tool. The computer program of the CNC machinery is, essentially, the set of coded commands that will dictate which motion or function the machine tool carries out in the production process. The program can be written manually by a trained programmer or can be created by using a more advanced programming software tool. The MCU is the computer that actually stores the program and executes the commands, and it is made up of a data processing unit and a control loops unit. This machine tool can be anything from a lathe to a mill depending upon what the specific manufacturing process calls for. The machine tool can be one of any variety of manufacturing tools, and is controlled by the previously mentioned components of the CNC machine.

When writing the program that the CNC machine is to execute, there are a variety of codes and programming languages, but there are universal standards. Most common CNC machines use a coded language called "G-Code." There are additional programs such CAM (computer aided manufacturing) software which can be utilized to simplify more complex programs. These CAM software systems can be purchased and run on your home or remote computer and later input into the machines main computer system, in order to continue to produce more complex codes at a more rapid rate by utilizing different software functions in relation to schematics and design. This CAM written program can then be loaded into a CNC system, which will directly input the program into the CNC data processor. However, this technique will be employed as more complex programs need to be written. As the more versatile the manufacturing industry gets, the more complex solutions need to be simplified. User friendly software as well as giving manufacturers other computer controls they need in their production process are very important factors to consider in the CNC manufacturing process.

As you can see, there are a variety of components involved in CNC machinery, with each of them serving a vital purpose during the manufacturing process. It is no wonder why a CNC machine can provide your business with a myriad of benefits that could not be attained using manual machines.

Inside each and every CNC machine there is some type of linear or rotational reference point to measure from, from which the part or product will be created. These dimensional references are called "axes", the well known XYZ coordinate measuring system. Just as in geometry, physics and other axes conventions that rely on the Cartesian coordinate system which was developed by Renee Descartes in early 17th century. The greater the number of axes of motion that a particular CNC machine is equipped with, the more complex functions and processes it can carry out. The axis' motions are either linear or rotary guided. A linear axis will travel in a straight line, usually depicted with an X, Y or Z to determine direction, while a rotary axis will travel in a circular motion around a fixed horizontal or vertical axis usually depicted with an A, B or C. The number of axes which are operated by the Computer Numerically Controlled system will play a large role in a machine's main purpose and capabilities.

Benefits of CNC Machinery

There is a vast array of benefits to using CNC machinery. Here are just a few of the advantages that CNC machinery can potentially bring to a business within the manufacturing industry today.

Engineers can utilize more advanced CNC equipped machines and design software to create commands that are able to manufacture products at a faster rate.

As some products require extreme precision and complex design, CNC machinery has the capability to manufacture those items, which in some cases could require higher skilled technicians.

Consistency is always a highlight maintained by CNC machinery. When working with CNC machines you can rely on the fact that a particular design that has been programmed into the computer will be manufactured to the same specifications. These machines have capabilities to maintain consistency to the thousandths and ten thousandths of an inch.

A business is able to operate a CNC machine around the clock all year, and will only have to shut it down periodically for maintenance and inspections. Usually CNC machines can run up to three shifts a day and continually produce parts.

Even machinists who are not highly skilled will be able to operate CNC machinery by using computer-assisted software as opposed to manual machines which typically require extensive training and experience to create complex parts. A number of mills and lathes usually require a skilled machinist not only to operate, but for precision, consistency and quality.

Labor costs can be greatly reduced, in that one person can program and run several CNC machines at once, with proper CNC training. Once a CNC machine is properly programmed to carry out its given task(s) it can, in some cases, be left with very little supervision.

In the event that a skilled engineer has a concept for a new product, changes to a current product or a more efficient way to carry out a task; they simply create a new program or change some code from the existing program for the changes to instantly create the new part.

CNC machines are able to produce various sized lots, as well as a variety of different products or pieces per machine, without having to have lengthy down times. In manual machines, it would often take quite some time to get the machine set up for the next work piece, therefore larger lots would typically be manufactured in order to be more cost and time efficient. However, when making changes to a part using CNC equipment, versatility is a key advantage.

CNC machinery also encompasses features like self diagnostic system checks that monitor the internal systems of the machines' computer and mechanisms. If anything should go wrong or not be working as effectively as it should be, the system will alert the operator and in some cases, specify the exact error and possibly provide solutions. These machines have become more responsive and informative with its feedback, which is important to assist with quality control.

Given that CNC machines minimize costs, maintain precision, constancy and efficiency, CNC machines are valuable additions to the manufacturing process. A common misconception about CNC machinery is that it is much more costly than their manual counterparts. Though this may have been true in the past, now, entry level or newer used CNC machines can be purchased for much less than the new manual machines counterpart. A person can find and buy a used CNC Machining Center (CNC Mill) for the same cost as a new manual mill. New and used CNC machines might have a slightly higher cost in some cases but they are still well worth the investment when comparing their production potential as a whole.

Chapter 2

CNC Machine Industry Past to Present

A Brief History of CNC

In order to properly understand the importance of CNC machinery it would be nice to take a look back to the birth of CNC machines, to get a clearer understating of its origins and see how the manufacturing industry has matured ever since. It is clear to see that the impact CNC has made in the manufacturing process worldwide has played a major role in a given countries economy and daily living. The manufacturing industry has grown by leaps and bounds with technological advancement and in large part due to integration of CNC.

CNC machinery can be traced back to the 1970's, thanks to the advent of the computer. However, NC (or numerical control) machines have been around since the 1940's. Though they were more rudimentary machines they did not have the integrated technology of the CNC machinery that exists today, they were still a significant advancement in the manufacturing technology of the era. They were made by using tools that were already in existence such as mills, lathes and presses, which were modified with motors that controlled their movements. Essentially, designated points were placed on punched tape that was fed commands into the main system to control the equipment.

Taking a look even further back into the history of CNC machinery, one will find that these machine tools were truly worked in conjunction with technology during the Industrial Revolution. In 1725 knitting machines that were controlled via punch cards were being utilized in England and music boxes that used the same mechanism were all the rage in various parts of the world. Even pianos were designed to follow this same fundamental principle, which used rolls of paper that were perforated in order to designate which piano key was to be autonomously played.

Between 1793 and 1825 Eli Whitney, inventor of the cotton gin, played a major role in the advent of interchangeable parts. Eli developed jigs and fixtures for milling machines as well contributed to the innovation of the manufacturing industry. But Eli was one of many people all over the world working on the "American System of Manufacturing" which used interchangeable parts and mechanization for production. Within that very same time period, more specialized metal machining tools began to be produced, such as a milling machines, lathes and other equipment geared specifically toward mass production.

Moving forward to the 1940's, John T. Parsons is credited with the advent of "Numerical Control." He was a machinist as well as a salesman at his family's machining company, and NC (Numerical Control) equipment was really created out of a necessity. Parsons was given a contract to build wooden stringers of rotor blades that were to be used for helicopters, and was seeking a new way to achieve this at greater speeds and with maximum efficiency. Despite repeated attempts at perfecting the production process, it wasn't until Parsons teamed up with Frank Stulen, who was the head of the Propeller Lab Rotary Ring Branch, that a successful system was developed for mass production using Numerical Control.

Working together, Parsons and Stulen thought to utilize the punched card principle in order to create a 200 point outline that could ultimately be used to manufacture a template for the stamping of metal stringers. This program was eventually brought to the shop floor of Parson's machining company. The machine operators used the "plunge cutting positioning method", wherein they would manually bring the cutting head to the designated spot and then lower the tool itself to cut. The only issues with this prototype were that it required a number of workers to operate it and the template that was produced had to be finished manually or by hand.

It was at this time when Parsons developed the idea of creating a fully automated manufacturing machine tool. He came to the conclusion that, if there were enough points on the outline itself, a worker would not be required to finish up the part. Upon hearing of the problems that faced the US Air Force regarding their new jet powered design, Parsons proposed to Lockheed (who held the contract for manufacturing from the Air Force at the time) that they use his newly invented automated mill to accomplish the task. Instead, Lockheed chose to continue on with their 5 axis templates, and failed to fix the problems that had plagued the project all along. As a result, Parsons was granted funding for the project.

Unfortunately, Parsons had problems of his own when it came time to build the machines that would need to be used during the manufacturing process. Rather than a polished and smooth outline being produced, the results were rough and still needed a person to clean them up at the end of the process. In his efforts to improve his design, Parsons turned to a team at MIT, who specialized in mechanical computing systems. MIT developed a way to eliminate the number of points that needed to be used in the outline to manufacture a smooth line to be cut. However, after the project officially ended in 1950, MIT acquired a milling machine company of their own and excluded Parsons.

After having been excluded by MIT, Parsons filed for a patent "Motor Controlled Apparatus for Positioning Machine Tool" in 1952, and was granted the patent in 1958. Licenses or sub-licenses were sold to IBM, Bendix, Fujitsu and General Electric. MIT, on the other hand, filed for a patent on "Numerical Control Servo-System" roughly three months after Parsons filed for his in 1952. It was when the servos were able to be controlled by APT (Automatic Programmed Tool) programming language that CNC saw another growth spurt. The key difference between MIT's machine and that which Parsons invented was that MIT used a 7 track punch tape, rather than a more simplistic punch card design. Three of these tracks controlled the axis, while the other four encoded information for the controls.

The MIT system made its debut in September of 1952. Though it was regarded as a breakthrough in manufacturing technology, it was also extremely complex and expensive. In all, there were a total of 250 vacuum tubes, 175 relays and a project bill of roughly $360,000. The key benefit of this system was that it was efficient, accurate as well as produced complex parts. These machines were used to manufacture a number of parts for aviation firms until around 1956.

The next focus in the evolution of numerical control equipment was the numerical controller. Concord Controls was a company that was largely made up of MIT engineers who left in 1955 to form a company of their own who developed the Numericord Controller. Similar to MIT's NC design but rather than using a punched tape to program the machine, the Numericord Controller used a magnetic tape reader. These machines were not as complex as the original machine designed by MIT and the magnetic tapes (which were produced by transferring punch card tapes to magnetic form) could be used on virtually any machine on the manufacturing floor.

Despite the fact that NC machines were becoming popular in some industries, many manufacturing companies were still hesitant to convert to NC machinery. This was because the time that was saved on the manufacturing floor was simply moved to the design of the punch card or magnetic tapes that were used. In essence, more office work was created even though production was higher. In fact, Parsons later stated that the Army actually had to purchase 120 NC machines and lease them out to manufacturers in order to make the concept of NC a more popular one.

CNC Changes the Manufacturing Industry

One of the main issues that plagued NC machinery was the fact that it did not have universal code. Instead, each manufacturer used its own language to program the machine's system. However, in 1956, the Air Force proposed that a generalized language be created (otherwise known as "g-code"), whereby a user could input a list of speeds and points into the system and the program would automatically generate a tape. And so CNC machinery was officially introduced to the world. In February of 1959, the APT programming language was announced at a press conference and a fully computer-controlled numerical control system was unveiled. The g-code was standardized by the Electronic Alliance Industry in the early 1960's.

The early 1960's also saw the development of CAD (Computer-Aided Drafting). One can trace CAD back to January of 1959 when both the Electronic Systems Laboratory and the Mechanical Engineering Department's Design Division of MIT held a series of meetings over several months that began the "Computer-Aided Design Project." Other companies such as General Motors were also attempting to computerize designs that were being created as well. Ultimately, it was a team at MIT who developed the first true CAD system, which led to surge in CAD start up firms in the 1970's. CAD, essentially, eliminated the need for paper drawings as well as draftsmen engineers. CNC machinery began to slowly but surely take the place of more antiquated machines that relied on older technologies.

Hydraulic tracers and manual machines began to be phased out in many sectors and the benefits of CNC machinery were now apparent to a number of manufacturers. Complex designs could now be produced with ease, quality and consistency. Fewer employees are needed to carry out the manufacturing process. Machines could quite quickly be programmed to carry out different tasks while the frequency of errors is greatly reduced. Another key reason why CNC machinery was becoming so widely used was that they were actually more economical for manufacturing companies. Not only were they less expensive than their earlier NC counterparts, but they reduced the company's payroll as well.

However, given that the economy within the Western world was experiencing slow growth, it was actually Germany that became a top supplier of CNC machinery in the 1970's. In fact, though the United States might have been credited with the invention of CNC machines and programs originally, the Germans were actually selling more CNC designs in 1979 than the Americans. Their reign was short lived though, as the Japanese took the "top producer" title in 1980. Many people have speculated that the United States CNC suppliers were ranking low in terms of sales because they were focused on higher end CNC applications rather than low-cost machinery. While the United States CNC industry had its sight set on providing machinery and programs to the aerospace industry, other countries were making designs that were selling rapidly in other sectors.

Shortly after this time period, CNC designers began to network CNC equipment together utilizing DNC (Direct Numerical

Control). DNC is basically a computer that can communicate directly to several different NC machines. This meant that an operator could, in essence, control a variety of different machines on the factory floor from one central computer. And, taking it one step further, some programs for certain manufacturing processes were too large to fit within the memory of certain CNC complex. DNC allowed for programs to be stored on another computer and sent directly to the machine.

It's important to mention that, even though CNC has greatly evolved since it first began, the oldest machines in the industry still utilize the punch tape system to control the motions of the equipment (though the tapes are now made of Mylar rather than paper). Some of the oldest NC equipment might have the ability to have backwards compatibility with programs that are already in place and allow for companies to utilize older machines on their manufacturing floor. Essentially, punch tapes are still in use because they save already existing manufacturing companies some time and some money in certain instances; however they are quickly dying out. Keeping the tapes means that companies would not have to take the time to convert their existing tapes into a different format. Also, they would not have to waste valuable resources to explore various other alternatives to a system that already was efficient and effective (as well as profitable). On the other hand, most companies have now replaced the punch tapes to a certain extent by using other forms of media, such as flash drives, floppy disks and local area networks.

Overall, one might say that CNC equipment (and its NC predecessor) is one of the most important innovations within the manufacturing industry. As CNC machinery has evolved into the machines that they are today, they are sturdier, faster and "smarter" than their earlier versions; it's safe to say that they are continually changing how the manufacturing industry evolves as well. CNC, by and large, allows for better quality manufacturing, more streamlined production processes, and consistency. The uses of CNC machinery are virtually limitless and throughout history we can see how it has changed the manufacturing process around the world.

CNC Economic Impact

Without question, CNC equipment has had a significant impact upon not just the economy of its birthplace, the United States, but on the world. CNC machinery has made it possible to make complex designs and has streamlined the production process as a whole. In fact, companies that subscribe to the philosophy of utilizing old, more antiquated machines on their manufacturing floors are running the risk of losing out to their competition and being phased out of the industry.

CNC machinery has drastically altered the way manufacturing companies of all varieties do business and has been a prevalent force in the economy of the manufacturing industry. And, as the manufacturing industry within any country is a key component of its micro-economy, CNC machinery has the potential to greatly affect a nation's economic status on a global scale. What I mean by status of manufactured goods refers to quality control of the manufactured goods in the USA, Germany, Japan, China, Mexico and other countries. Some countries make better quality parts than others.

Here are a few of the ways in which CNC machines have impacted the economy in some way, shape, or form:

- Reduced Production Costs

CNC machinery allowed for companies to produce goods at a lower cost. This was mainly due to the fact that companies did not have to employ as many workers to carry out the manufacturing process, as the machinery and associated programs took on a great deal of the work required. Also, there was a decrease in wasted raw materials, as the machinery was able to produce the items with more efficiency. As a result, items could be produced at a lower cost, with the savings being passed on to the consumer.

- Decreased Production Times

Manufacturing companies are able to produce products at a more rapid rate without having to move machines or parts around on a constant basis. Therefore, items could be mass produced, which lowered the cost as well as the man power that was need to produce the item originally.

- Consumer Saving

Companies who now rely upon CNC machinery can now produce items that are high in quality, less costly to make, and can be produced more quickly than ever thought possible before the advent of CNC machines. Therefore, their profits were greatly increased, and the general public could now purchase items at a much lower cost as well.

- Improved Quality

CNC machines are precise and, thus, can produce items with improved quality, without sacrificing speed. CNC machinery and their programs leave little room for error, which lead to a higher quality produced product and very few discarded items because of defects in design or production. And, with the constant evolution of CNC machinery and design and the creation of new and improved programming capabilities, the quality and production speed of manufactured products is sure to only get better over time.

When you take a closer look at CNC machinery, you will also discover that virtually every sector which it has touched has seen a boom in their industrial economy and a significant improvement of their overall production speed, quality and a decrease in production costs. From the furniture making industry, to companies that provide highly complex parts for computer systems or medical parts, CNC equipment is significant in nearly every sector of manufacturing.

The automotive industry is one such sector that has seen tremendous growth since NC and CNC machinery has been introduced. While the first cars were produced with a myriad of laborers who worked on an assembly line, each performing the same receptive task throughout different stages of the manufacturing process, autos are now produced using sophisticated CNC machines (aided by skilled operators) that are able to manufacture automobiles quickly, at a lower cost and with more precision. This leads to a decrease in overall production costs and an increase in product quality which is passed on to the consumer. In addition, many more cars are able to be made each year, in order to keep up with the constant global demand. Thus, if CNC machines weren't around today, the delicate balance of supply and demand that keeps many such manufacturing companies profitable would be greatly disturbed.

Chapter 3

Basic Machine Details

Machine Basics

If you are planning on utilizing CNC machinery in your business or even for personal use, then getting familiar with the basics of CNC Machinery is highly recommended. CNC programmers, operators and owners most likely will have a deeper insight of components, functions and controls that are typically involved in most CNC machines. For this, we will cover three common types of machines and explain some components, functions and controls.

CNC machines are machines such as mills, lathes, presses and many other manufacturing tools equipped with a computer and additional components that will run the programmed commands for a specific operation. CNC mills are known as Machining Centers, CNC lathes are known as Turning Centers and CNC Press Brakes are presses that "brake" or bend sheet metal to name a few.

Let us consider a mill with the drill in position to drill into a block of metal or wood. Then you proceed to drill 254mm deep into the block at a specific point on the block. Of course the block has a length, width and height like 300mm x 400mm x 500mm. As a rough example we will call the length the x-axis, the width the y-axis and the height the z-axis. In real life applications there is a zero reference point or "home" position that the machine will start from as it moves along a certain axis to cut. Now there are several key components to a mill. The "drill" has many parts in itself but the main part of the drill is known as a spindle. The spindle is the part that spins and in the spindle is where the cutting tool or "drill bit" would be interchanged as needed or required. These spindle speed and cutting tool changes are to provide for a more efficient and precise cutting operation. An operator can control the direction of the drill as it moves along the X, Y or Z-axis, the speed at which the spindle rotates as well as the diameter of the "drill bit" inside the spindle along with special features sometimes called options such as an Automatic Tool Changer or additional axis.

The lathe works in a similar manner with the exception that the part being cut is usually a bar or rod. A lathe holds a cylindrical work piece on one or both ends such as a 100mm long bar with a 15mm diameter. The cutting tool is then gradually passed along the surface of the rotating part (bar or rod). With the bar spinning, per a given length of bar, the cutting tool can cut away the material to create vast array of products such as screws, bolts and needles to name a few. On a lathe, the spindle holding the part is known as a chuck. The chuck can vary in size and is required to clamp or hold the part (bar or rod) securely. The lathe has a directional reverence to measure and cut by just as a mill does, usually the cutting tool on a lathe will cut on X and Z axes, which is the industry standard.

Another machine we can look at is the press or press brake which will press or bend material such as bending or stamping a shape into a piece of sheet metal. Press Brakes specifically bend by the shaping of sheet metal, straining the metal around a linear axis. The bending or "braking" operation compresses one side of the bend and stretches the other side of a piece a sheet metal. Press brakes can be operated by a computer to bend a piece of sheet metal to a specific angle to creating multiple bends with precision and accuracy. Press brakes are mostly chosen by tonnage required bending a given thickness and length per a given type of material such as mild steel or iron. Usually there is a tonnage chart that will show material thickness and die opening that corresponds to the tonnage required to bend.

As an example to bend a 1000mm x 2000mm, 100mm thick piece of mild steel sheet metal might take 60 tons even more depending on the tensile strength of material. The piece of sheet metal would be inserted between the machines housing side frames that the piece will fit through. In this case we need to fit in the 1000mm length side of the sheet metal. Usually there is plenty of space for the 100mm thick metal to fit in, the "Stroke" should be larger than the piece of sheet metal desired to bend. Through CNC a press brake can bend to an exact angle and then CNC robots can transfer the material to storage rack and place a new sheet of material to be processed without an operator present. CNC can create a completely automated production system.

Other machines might cut by lasers, others might cut with fire like plasma cutting machines and some even cut with water such as waterjets. More and more CNC manufacturing tools are being created for the CNC manufacturing industry as it continues to grow. Although the industry continues to evolve there are key components that the industries rely on such as capabilities, speed and accuracy. Each type of CNC machine is designed with those key components in mind. Each machine is developed to enhance manufacturing capabilities and execute specific functions with maximum efficiency.

CNC Motion Controls, Programmable Functions and Options

As previously discussed in this book, CNC machinery relies upon manufacturing tools that carry out the programmable functions that have been inputted. Typically, A, B and C are used for rotary axes and X, Y and Z are utilized for linear axes. You should also take note of the directions of each axis, such as whether they are plus or minus, and the reference point for each axis. This reference point can also be called the "home position" or the "zero position." It is necessary to know this exact point, so that you are able to provide the control with an accurate point of reference when inputting the program.

Every CNC machine also has specific motion types that it is capable of. Three common types of motion that are found in CNC equipment are: rapid motion, straight line motion and circular motion. Rapid motion is to shorten the length of time in which the equipment is inactive while machining. It is utilized while the machine is operating at its fastest rate, and is sometimes referred to as "positioning." Straight line motion is used to specify the feed rate, as well as to make straight line movements. This type of CNC machinery motion is also known as "interpolation." Circular motion, which is also referred to as "circular interpolation", is used to create circular motions while machining, as the name suggests. Some CNC equipment may even feature a helical motion, or a variety of other motions that enhance the functionality of the CNC machinery

Aside from the motion controls, there are a wide range of programmable functions that can help to make the operation of any

CNC machine more manageable. While some older or low cost CNC machine tools may require you to make changes to the process manually, newer sophisticated equipment will allow for you to program those functions. These CNC machines can carry out functions if left on their own, and don't require the operator to be present after the machine has begun.

The codes for these basic functions will widely vary from machine to machine, so it's best if you consult with your owner's manual to find out which particular codes are required for your CNC equipment. Spindle control, automatic tool changer, coolant control, automatic pallet changer, and part changes are all components or functions of a CNC machine that can be controlled by the computer program. So, as you can see, it's important that you learn the codes for these basic functions in order to sufficiently operate your CNC equipment and to keep it in proper working order.

There are a variety of functions and features which may come standard with your CNC machine or are available for additional costs known as options. These additional options or accessories can enhance the functionality of your CNC machinery and make them even more efficient and productive. Though the choice of options will vary greatly based upon your particular CNC machine manufacturer, the options that are included with your new CNC machine should be noted in your purchase agreement, owner's manual, or look online to see which options may help you to get the most out of your CNC equipment.

Examples of commonly used options and accessories for CNC machinery include: automatic pallet changers, bar feeders, probing systems, lighting, safety features, and other miscellaneous and adaptive control systems. However, there are options available for virtually every popular CNC machine that is manufactured today. There are systems that can be put in place to aid in the automation of your existing CNC machine. If you are finding it difficult to find options and accessories that are ideal for your CNC machinery uses, it may be a good idea to contact the manufacturer directly to see if they may be able to make recommendations that may enhance your CNC machine tool.

Sample Specification Sheet

Every CNC machine and tool will have its own set of specific components. If you are planning on purchasing a CNC machine, have already acquired one, or are going to be operating or programming a CNC machine, then there are key components that you should know or clarify before using the machine. The CNC machine manual should be able to provide you with the information required to determine if the machine is right for you and is capable of producing the part and results you require. Whether large print or on a small metal plate attached to the back of the machine, there will be the make, model and serial along with the age. Knowing the make, model, age and serial number of the machine will allow you to find additional information pertaining to the machine online as well as directly from the OEM (Original Equipment Manufacturer).

See the following Sample Specification Sheets for specifications, options and detailed examples of what to expect when looking into CNC equipment information. These are just examples; there are far more specifications and details vital to CNC machinery than the ones listed here:

Machining Centers (CNC Mill):

ExampMach MC-1000 Machining Center, New: 2010 (age machine was built)

Specifications (key details):

Travels: 300mm x 400mm x 500mm (x-axis: 300mm, y-axis: 400mm, z-axis: 500mm)
Spindle speed: 4500rpms
Motor: 15 hp

Equipped With:

Exacomp MC-1 CNC control (software preference and required)
Thru Spindle Coolant (option)
ATC (Auto Tool Changer): 20 Tools
All manuals available

Basic Machine Information:

OEM: ExampMach
Model: MC-1000
Machine Type: Machining Center
Age machine Built: 2010

Additional specification sheet information:

- Travels determine the distance the table or spindle can move to cut a part per size of part.

- Spindle speed is measured in revolutions per minute.

- Motor specifications in addition to hp (horse power) are crucial to machine operations. There are requirements and specifications such as voltage, wattage, amps and other properties that are that just as important.

Equipped With are additional features or options that will be included with the machine:

- CNC control is very important as it is the computer and language that the programmer will use to operate the machine. The control will have a name and model or version.

- Thru Spindle Coolant is an option on some machines that will drive coolant through the spindle rather than pouring coolant on the spindle as the spindle needs to be cooled.

- Auto Tool Changer has options available on most machines; in this case it has up to 20 slots to hold tools also known as tool holders which hold the drill bit.

- All manuals available are added to "used" equipment purchases due to the fact that the owners may not have all the manuals available. It's a statement that is important to note since you might want the operating manual as a reference or in the event that the machine malfunctions.

Turning Center (CNC Lathe):

ExampMach TC-1000 Turning Center, New: 2010 (age machine was built)

Specifications (key details):

Chuck:	152mm
Maximum Swing:	304mm
Maximum Machining Diameter:	203mm
Maximum Machining Length:	381mm

Equipped With:

Exacomp TC-1 CNC control (software preference and required)

Tailstock

All manuals available

Basic Machine Information:

Model: TC-1000
Machine Type: Turning Center

Additional specification sheet information:

- Chuck is the devices that will clamp and hold a cylindrical work piece as it rotates on the lathe or turning center. The "jaws", which are the parts of the chuck that actually clamp or hold the work piece usually called "3-jaw" or" 4-jaw" chucks.

- Maximum Swing determines the maximum diameter of bar that can spin under the machines capabilities.

- Maximum Machining Diameter which is a slightly shorter length than the maximum swing due the

- Cutting tool and length required to fit between that and the part.

- Maximum Machining Length also known as the (DBC) Distance Between Centers, the maximum length of surface can be cut along the work piece.

Equipped With are additional features or options that will be included with the machine:

- Tailstock is an accessory used to support the backend of longer work pieces opposite end of the chuck.

CNC Press Brakes

ExampMach PB-1000 Press Brake, New: 2010 (age machine was built)

Specifications (key details):

Maximum Tonnage:	1000 Tons
Maximum Bed Length:	3650mm
Distance Between Housing Frame:	3040mm

Equipped With:

Exacomp PB-1 CNC control (software preference and required)
Crowning Compensator
All manuals available

Basic Machine Information:

Model: PB-1000
Machine Type: Press Brake

Additional specification sheet information:

- Maximum Tonnage is the maximum tonnage the press can apply. Keep in mind the press brake can be adjusted to apply less tonnage.

- Maximum Bed Length is the length of the bed that can bend before the bed frame comes to be a factor.

- Distance between housing is the distance a work piece can fit into the press brake between the frame for additional bending capabilities.

Equipped With are additional features or options that will be included with the machine:

- Crowning Compensator is an option to correction for deflection. Common in press brakes with longer frames and bed lengths, the machine frames tend to deflect during the bending process. The bending angle is not constant over the entire length of the bend. And as a result, the bend will be slightly inaccurate across the length of the bend, especially towards the center of the press brake which needs to be corrected.

Knowing key components about your new or used CNC machine will enable for you to become comfortable with the basic functions of the machine, as well as to know its capabilities and its limitations. Of course there are also many online forums, websites and reviews that can be used as a resource for additional machine information. Again, it's highly recommended that you thoroughly read through the owner/operator manuals before using your machine, in order to learn as much as you can about its components, functions and capabilities.

Chapter 4

Introduction to CNC Machines

Major CNC Machine Types

Here is a short list to give you additional insight as to CNC machine types, some that you might already know.

- Machining Centers

A Machining Center or "CNC milling machine" is used in the production of solid materials, and can carry out a wide range of necessary functions that are required during the manufacturing process in a variety of industries. From functions as simple as drilling to those that are more complex, such as contouring. They are usually classified as being either horizontal or vertical machining centers, which has to do with the positioning of the spindle. There are benefits associated with each type of milling machine that you could look into depending on your goals. There are also a number of mill machine variants, ranging from bed mills to floor mills, each serving a particular purpose.

For example, as with horizontal machining centers and vertical machining centers, spindle position is the key factor. If you are planning on manufacturing larger items, whether they are longer, heavier and/or positioning is a concern, you could benefit more from one type of or the other.

When you begin to conduct your research on which CNC machine you would like to ultimately purchase, you will probably notice that a variety of machines are referred to as being either vertical or horizontal CNC machines. The jobs that you are going to be carrying out on the machine, as well as the size and shape of the parts that you are going to be manufacturing, should greatly determine which CNC machine orientation you opt for.

- Turning Centers

CNC Turning Centers and other tools similar like boring machines use a single point cutting tool, which is moved parallel to rotation axis of the machine, to manufacture a part. The part is typically rotated during the process. There are a variety of different operations that a CNC turning center and boring machine can carry out with the correct attachments or modifications, including: spherical generation tapered turning, facing, grooving, boring, drilling, and threading of parts.

- CNC Router

CNC routers are a cutting tool common in manufacturing and are commonly used to produce larger items, such as cutting wood or even sheet metal. However, more sophisticated CNC routers that carry out more complex tasks may have multi-axis cutting heads. A router can increase production speeds, minimize errors and decrease the amount of waste produced while manufacturing.

- CNC Waterjet

Waterjets are becoming more common and are used primarily on sheet metal and other composite material where water is used to cut the material. With enough water pressure, water is able to cut through stone and metal. In many cases there is sand or granite mixed in the water to assist the cutting process as well as multi-axis cutting heads, which tilt and broaden the machines capabilities. These machines are used for architectural and landscape designs to medical tools to common parts within the products of your home.

- CNC Plasma Cutter

A CNC plasma cutter has the ability to cut steel and other metals by using a plasma torch. A computer program produces the amperage for the torches plasma cuts, rather than a worker, which can reduce the safety risks and ensure cleaner lines. A CNC plasma cutter is ideally suited for tasks that require cutting thicker materials. However, new or thinner plasma cutter nozzles that produce a finer arc are now widely being used for jobs that involve the cutting of thinner materials. While CNC plasma cutters used to be quite expensive, heavy and large, they can now be rather affordable and portable pieces of equipment especially in the used market. As it is with all CNC machinery, the design of the plasma cutter is constantly evolving, and newer innovations are being used to better them on an ongoing basis.

- CNC Grinder

A CNC Grinder is used to grind away material to the microns over the surface of a part. CNC grinders have the capability to carry out a variety of operations (i.e. surface or cylindrical grinding tasks). A CNC grinder typically uses a grinding wheel. CNC grinding tools enhance productivity and allows for longer periods of automation, wherein an operator can focus on other tasks on the factory or shop floor. They can also be modified into what is known as a "High Speed Precision Grinding", whereby they can manufacture highly complex components, with precision, quickly and efficiently. The aerospace and medical industries greatly benefit from grinders, as the materials they use for parts they need are often hard or exotic. The modern day CNC grinder can easily manage more difficult raw materials. Several industries rely on grinding parts on every detail of their product. Some product certifications require that all parts in a product be grinded to the microns and to perfection. The detail and precision of CNC grinding machines go into making airplane parts, prosthetics, watch parts, automotive parts and more.

- CNC Laser Cutter

CNC laser cutting machines have some of the most advanced components in the CNC equipment industry. Certain CNC laser cutters are cutting not only steel, but aluminum, copper, brass and other metals. Like routers and waterjets these machines usually cut along the x and y axis. The laser cutting machines have multi-axis heads which pivot, to add more capabilities. The wattage required will usually be determined by the thickness and material of the sheet metal intending to cut. Although these mainly cut sheet metal there are many other ways CNC laser equipment is being utilized such as cutting pipes, tubes and engraving. While still being able to accurately carry out jobs that deal with heavy metals, you must equip the machine with different laser strengths for various materials and cutting speeds.

- CNC Engraving Tools

CNC engraving tools can be used to engrave a variety of materials. Glass, wood, metal, and plastic are often the most common mediums used, however CNC engraving machines can permanently etch designs or lettering on to virtually any material. Small, medium and large engravers are available via CNC engraving tool distributors and specialty engravers. CNC engraving machines can also be fitted with different bits, such as diamond or stone, to tackle heavy duty jobs, such as steel engraving projects.

- Multi-Axis Machines

These more complex CNC machines are used to manufacture parts, using more axes of motion in their machines. Typically, these features are utilized on machining centers, turning centers, Swiss screw machines, waterjets and laser cutting machines, to produce various cuts on various materials to produce various parts. For example, CNC 4-Axis machining centers have the capability of making more complex parts than the standard 3-Axis machining centers simply with an additional axis to rotate the part. The work that must be performed to rotate the part is greatly reduced. You can have a multiple axes on many machines such as turning centers or Swiss screw machine which could have 10 or 11 axes to produce extremely complex parts.

Major CNC Machine Brands

Given that CNC equipment has become very popular in recent decades, you will find that there are a plethora of manufacturers to choose from, all of whom boast that they are the "premier source for all of your CNC machinery needs." However, what you must be aware of before clicking that buy button or handing over your preferred method of payment is the reputation of the manufacturer and the respectability of their brand. You will undoubtedly want the best quality and value without sacrificing required components or features. Here is a short list of major CNC machine brands that may be worth investigating when you are ready to purchase your new CNC machinery.

- Mori Seiki

This CNC machinery manufacturer has been around since 1948, and currently owns a variety of subsidiary brands like Hitachi Seiki. They are known for their quality turning centers, machining centers and other CNC machine tools. They offer a wide range of vertical and horizontal machining centers, multi-axis CNC machines and CNC turning centers and other types of equipment. They have won numerous awards for their machining centers, and hold a respected place in the global marketplace with their high performance CNC machinery.

- Haas Automation

This CNC machinery manufacturer specializes in vertical machines, and has been making them since 1988. They have over 70 different types of vertical CNC machines, ranging from a small office mill, to their high performance machining centers. Their Haas TM series Tool room mills provide an inexpensive way to transition from manual to CNC machinery, and even come standard with an intuitive programming system that enables those without any knowledge of G-code to set up and operate the equipment quite easily. Haas Automation is considered to be one of the leading manufacturers of CNC machinery in North America, and they are renowned for their low cost, easy to use equipment. Also, they have a CNC machine that will meet virtually every need that manufacturing businesses, of all types and sizes, may have.

- Toyoda

Toyoda (not to be confused with the auto manufacturer) offers horizontal and vertical machining centers, bridge and gantry, grinders, automation equipment. Their high-performance machinery is utilized in the auto, energy, aerospace, medical and heavy equipment industries. They also offer a wide array of services (ranging from spindle reworking to engineering).

- Mazak

This particular CNC machine manufacturer is well known for its blend of quality and state of the art technology. They currently sell turning centers, multi-tasking machine centers, 5-axis machines, horizontal and vertical machines, and their NEXUS series of CNC machines. Their NEXUS series, as most of their equipment, combines performance, value and technology to provide shops of all sizes with the ideal CNC equipment that they require carrying out a multitude of manufacturing tasks quickly and efficiently. They supply CNC machinery for the nearly every industry such as auto, medical, aerospace, energy and mass production industries, just to name a few.

- Okuma

Known for their "world class machine tools", Okuma is known for their high quality machinery and their attention to customer service. They offer a tech support hotline and access to replacement parts round the clock. They offer double column machines, grinders, 5-axis machines, pallet systems, and vertical/horizontal machining centers, as well as turning centers.

- Hyundai-WIA

From turning centers to lathes and multi-tasking centers, Hyundai WIA has a variety of CNC machinery to meet the needs of a wide range of manufacturing shops. They have been around since 1977, and are currently the top producer of precision machining tools in Korea, a title that is has held for a while.

- Citizen Machinery Co., LTD

They are known for their excellent for high-precision, high-speed machining and known in the industry for its ease of use. They provide applications for individual component machining optimized for the customer's specific needs, equipped with the latest network technologies in addition to CNC control technologies that make high productivity possible. From basic turning centers to advanced high-precision, high-speed machining, Citizen CNC machines have been known for their reliability, accuracy and speed.

This short list of equipment types and brands by no means represents the industry as a whole. Take a further look on and offline into the many Original Equipment Manufacturers (OEM's). You will find several OEM's that have stood the test of time and produced great CNC machine tools which have shaped our way of living. They all also offer service and parts support, as well as application support for those who need help with automation. You are able to reach them via their site or even by phone if you need assistance before or after your CNC machine purchase, which gives those who are looking for extended customer support some added peace of mind.

Chapter 5

Machine Condition and Inspections

There is a vast array of CNC machines in every sector of manufacturing. When discussing the various types of CNC machinery, it's important to note that these machines fall into many categories like buying new or used to reconditioned, retrofitted and custom built machines.

New and Used Equipment

Whether looking to buy new or used each vary in benefits. Buying a new CNC machine most likely will come with a warranty as well as the reassurance knowing parts, services and support will be available. The warranty, usually 2 or 3 years, will cover manufacturer defects and include discounts on services and parts that the machine might need. Another benefit of buying a new machine is being able to take advantage of the latest and greatest in technology. For precision and speed, as well as options available, is huge advantage of new equipment. These key features help increase productivity and make for a more efficient production process.

Buying used equipment has its advantages as well. Saving costs is a highlight of buying used equipment. Older pieces of CNC equipment are more than capable of producing excellent parts from mass production to prototype.

If you are considering switching from manual machines to CNC machinery, however, it's highly recommended that you purchase used CNC machinery. Quality pre-owned CNC machines are significantly discounted and can give you the opportunity to transition to all of the benefits that advanced CNC technology has to offer without having to spend a fortune.

Used equipment, especially when maintained properly, can have a long life of production. When buying a used piece of equipment; there is roughly a 30% percent price difference, or more, that you would save compared to buying a new. When you buy a used CNC machine, the service, support and parts still might be available and up to date. Be sure to research prior to purchase. When buying used equipment it is vital that an inspection of the machine is done prior to purchase. Inspections are important, especially in the case that support, service and parts are hard to come by due to age or condition of the machine. An inspection will give you an abundance of information as to the machine's condition.

Retrofit, Refurbished and Reconditioned

There are a number of benefits of transitioning from manual to CNC machinery in your machine shop or factory. As other types of industries are beginning to see the many benefits that CNC machinery can bring, an increasing number of machines are being retrofitted, refurbished and reconditioned with CNC technology in order to get the maximum effectiveness of their tools.

Though you can purchase machines that already have CNC controls built in, you may also choose to purchase retrofitted CNC equipment or retrofit your existing mill or lathe to include CNC technology. This will typically require that you make some modifications of key components that were built into the machine in order for it to be easily used by a human, such as motion control and motor speed and replacing those components with CNC associated parts (i.e. motors, mounts, wiring and computer hardware).

There are a number of advantages to purchasing a refurbished or reconditioned CNC machines as well. They are less expensive than new machines, yet you will typically find them in the price range between new and used. Refurbished and reconditioned CNC machines will usually have new parts that replaced any worn ones, thoroughly cleaned, and all of its cosmetic issues have been resolved prior to resale. If you can find a high-quality refurbished CNC machine where the motor, spindle, etc has been replaced or brought to like new condition is typically ideal, given that the life of the machine greatly depends upon the age and use. Regardless of buying any used, retrofitted, refurbished or reconditioned machine it is always good practice to inspect the machine prior to purchase. See how the machine has been maintained and/or see it production if possible to ensure that the machine will have a long production life.

Importance of Inspections

There are a number of benefits associated with having a CNC machine inspected and regularly maintained, which include detecting any problems that may lead to more costly repairs in the future as well as repairing any current issues that may decrease productivity or profitability.

Whether you are a CNC machinery owner who would like to make sure that your CNC machinery is proper working order, or are someone who is planning on purchasing a used CNC machine, carrying out a CNC inspection is highly recommended. A CNC inspection that is carried out by a professional, who has experience in dealing with CNC machinery and knows just what to look for, can provide you with a diagnostic of most any issues that need to be addressed on the CNC machine, as well as recommendations that may need to be made in order to keep it working efficiently.

There are a variety of reasons why you should have your CNC machinery inspected. Most commonly, a CNC machine inspection are for those who are purchasing CNC machines other than "new" or machine shop owners who are overdue for a routine maintenance inspection. Unfortunately, there are many CNC machinery owners who simply forget or neglect to have their CNC equipment regularly maintained and inspected, or do not feel that it is a worthwhile investment, but it is. Inspecting or hiring a professional CNC inspector to thoroughly inspect key components of your CNC machine is one of the most important and business-savvy things that you can do as a CNC machinery owner. Spending a minimal fee now to have your machine inspected can lead to thousands of dollars saved down the line. The more you know about the machine the more you will save.

Some of the basic components of your machine that will be checked during an inspection include: the spindle, bearings and ball screws, the machine's axes, tips, lubrication, various parts inside and outside of the machine and in some cases, how well programs are executed by the machine (i.e. if any unusual sounds are made while performing a programmed task), amongst other things. If an inspection reveals any concerns, addressing them early can lead to tremendous savings down the road, and can improve the safety of your machine, as well as productivity.

Below you will find a short detailed list of components that should be evaluated during a typical CNC machinery inspection:

- Parts/Functions: Spindle, gears, axes (including X, Y and Z), rotary, indexer, spindle run out, pallet operation, RPMS, repeatability, motors, bearings, ball screws, ways and linear guides, tips, heads, seals and more amongst a number of others.

- Repair/Damage: Detection of rust corrosion, fluid leaks, defective oil pans, damage to the inside or outside of the machine, and damage to the frame of the machine. This is crucial, as many parts can be costly to replace or repair if they are not taken care of properly.

- Other Parts: Options and accessories like turret, tail stock, foot switch, tool eye sensor, hydraulic oil, bar feed, coolant pump, coolant fan, live tooling, chip conveyor, power supply, chillers, heads, tool holders and any other components and functions that may be equipped on the machine.

- Overall Condition: Dents or dings in the machine, way covers, abnormal sounds while in operation, electrical cabinet, transformers, doors, windows and glass, fans, and the overall functionality of the machine in its current condition, while running under power preferably.

As you can see, CNC equipment inspections can detect a variety of problems that could have been prevented and will provide you with a thorough analysis of the overall condition of your CNC equipment. Therefore, choosing an experienced and qualified CNC inspection service provider is essential, as doing so will give you the peace of mind of knowing that your machine is good working order and all issues have been addressed.

Chapter 6

The Importance of CNC Machine Maintenance

In order to ensure that your CNC machinery works at optimal levels, as well as to extend the maximum life, it's highly recommended that regular maintenance and inspections are done. Given that you have most likely invested a significant amount of money to obtain a piece of CNC equipment, make sure that you have the machine thoroughly checked over. This chapter will discuss the basics of CNC maintenance, as well as the importance of having your CNC machinery inspected regularly.

Maximum Life of CNC Machinery

The maximum life expectancy of CNC machinery greatly depends upon which particular CNC machine you have purchased, and whether or not you have bought a new or used piece of equipment, of course. Also, keep in mind that having your machine regularly maintained as well as having inspections conducted often is a key factor in extending the life of your CNC tool.

Generally speaking, it's not unheard of to see a 30 year old CNC machine still in operation on the factory floor. However, the maximum life expectancy of any CNC machine is dictated by its control, motor and parts. Thus, when we speak of its maximum life, we are really talking about the life of the motor and computer software updates. With a CNC machine receiving proper maintenance and inspections, you could very well have that same CNC machine in your factory or shop for decades to come.

When you purchase your CNC machine, make sure to find out how long the motor is supposed to last. Usually this life expectancy will be given in hours. For example, a machine has a life expectancy of 50,000 hours, and then you can expect it to last for 25 years if you run it for 8 hours continuously, 5 days a week, with proper maintenance. Therefore, if you have taken good care of your CNC machine and have made sure that you have carried out proper preventative machinery maintenance, you may find that your CNC machine will last considerably long.

Another factor which typically determines the maximum life of a CNC machine is the quality of its design and how it was manufactured. For instance, if you opt for a lower end machine that only costs a fraction of what some higher end CNC machines may cost, then you may discover that you may only get 5,000 hours out of your CNC equipment. However, higher end machines may stipulate that they will run for 20,000 hours, which will account for a lot more production time vs. down time throughout the years, especially if you perform 40 hours of machining every week. Once again, it's important to mention that even lower end machines can last past their life expectancy if you take proper care of them.

Basics of Preventative CNC Machinery Maintenance

There are a number of factors that are involved in CNC machinery maintenance. From cleaning to repairs to regular inspections, it's recommended that you treat your CNC machine like the valuable "worker" that it is and make sure that every part of it is well cared for, especially if you want it to perform at its best.

Here are just a few of the factors that are involved in the CNC machine maintenance process:

• Cleaning the outside of the CNC equipment

Many CNC operators and owners may not consider the cleaning the outside of the CNC equipment to be a vital part of its maintenance to extend maximum life, but wiping off any excess coolant, water or chips of materials can help to prevent damage to the equipment and remove any safety hazards.

- Cleaning the inside of the CNC machine

The inside of the CNC machine must be kept clean at all times, but especially after a shift has finished. Chip buildup can actually cause a coolant blockage, which can result in heat damage due to low coolant pressure. It's also important to always make sure that you brush away any chips of material that may cause damage to the tools or prevent them from carrying out their work.

- Lubrication of the CNC parts

You should always ensure that your oil reservoir is filled and that the machine is well lubricated before use. It's always a good idea to warm up the machine after it has been turned off for any period of time, to enable the oil to properly circulate through the CNC machine's various parts before making any cuts.

Benefits of Regular CNC Machinery Maintenance

Aside from the obvious benefit of ensuring that your machine's life expectancy is extended as a result of regular maintenance and inspection, there are also other key benefits to properly caring for your CNC equipment. First and foremost, regular maintenance can increase the functionality and performance of your CNC machine. If you do your part to clean and inspect the machine regularly, then your equipment is going to run at optimal levels, as opposed to being slowed or shut down for repair work often. Thus, your CNC machinery will be productive and efficient, which results in less operational costs and significantly increased profits.

Next, regular maintenance and inspection can add to the overall value of your CNC machine or help it to retain its current value. If you should choose to sell your CNC equipment down the line, you will be able to sell it for much more if it is in good working order and there are no major issues with it. Spending just a little on maintenance, new parts, and inspections in the beginning can end up being a great return on your minimal investment. Also, you have to keep in mind that properly working machines will increase the value of your business as well.

Last, but certainly not least, getting your CNC machine regularly inspected and carrying out routine maintenance can ensure that the equipment is as safe as possible. Given that clogs, leaks, or broken parts of the machine can lead to certain hazards, having it serviced regularly can potentially protect your workers or your facility from certain dangers, such as fire or injury. Keep in mind that, even though these machines are more technologically advanced than their predecessors, they are still capable of harming operators if they are not respected, properly cared for or do not meet safety regulations as a result of lapsed inspections.

Chapter 7

The Future of CNC

There is no limit as to what CNC machines will be capable of as well as where CNC machines will be in the future. As Charles Benjamin Maginnis V, a CNC Machine Specialist said, "We have only scratched the surface of CNC machining."

There are a few concepts that could be mentioned or that might be of some significance in determining the future of CNC equipment. These machines are getting faster, more advanced, more precise as well as more efficient. As computer integration continues to be a key component of CNC equipment, we will be able to experience all the features of advanced technology. Engineers of all sorts will be able to take part in the innovation and growth of the CNC machine industry.

Technological Advancement

High speed, high precision, user friendly, energy efficient low cost high production machines are on the horizon. Hybrid CNC machines and improved CNC controls are all speculated to be in development as CNC technology is continually evolving. As newer CNC machines are being designed and built to include updated technologies, older high quality CNC machines are being sold as pre-owned models. Therefore, those who wish to enter into the CNC manufacturing business will have the opportunity to do so, even if they do not have the budget to purchase a new CNC machine. The manufacturing industry as a whole is constantly exploring new heights. There are several manufacturing businesses currently running CNC today and that number is only rising as business owners begin to see the many benefits of having CNC machinery in their facility.

Different Directions

Creation and innovation, especially in the manufacturing industry can mean different things to different people. Assisting the progress of CNC machines can take many shapes and forms. Not only in the manufacturing industry, but also in our personal life, computer controlled equipment has played a major roll. From parts used in our vehicles, parts used as home goods, to the parts that help make life more productive, computer-aided technology will continue to lead the way. These technological enhancements help the manufacturing industry be more productive and efficient. It goes without saying that these same applications can interact with the equipment and devices used in our everyday lives. Computer controlled equipment can shape the way we live and interact with equipment and devices. The possibilities are endless.

As technology advances, we have seen many concepts being adopted by the CNC industry. Virtual reality is one of those concepts. We have seen prototypes for manufactures and engineers to modify or control parts in a 3-dimensional virtual reality world, just as if you were working on the part right in front of the actual machine. Soon you will be able to put on 3D goggles or glasses, as well as other attachments like gloves and other accessories, which will give you full access to machining with safety, comfort and convenience. Once the prototype is complete in the virtual reality program, the user, in theory, would be able to input the program in an actual machine and end with the same result yielding an actual sample part.

As we know most machines and electronics are controlled by a computer-aided device. CNC machines primarily concern manufacturing equipment with specific computer languages and operations. But as programming and codes become more universal, more machines and devices will be fitted with computers controls. From machines manufacturing in the home, to mobile manufacturing equipment, computer controlled machines will continue to saturate the globe.

Let's look at the Internet of Things, a slightly different variation of computer controlled equipment, but similar in concept. The concept is controlling equipment by computer-aided technology such as Bluetooth, Wi-Fi and the Internet. The Internet of Things (IoT) is a network of physical objects or "things" embedded with electronics, software, sensors and connectivity. They enable it to be a better value and service through data processing with the manufacturer and/or operator using other connected devices using computers.

Obviously there would be too much to cover regarding the future of CNC, the equipment and industry. The possibilities are endless; almost any piece of manufacturing equipment can be turned into a computer controlled machine. Computer controlled machines are a part of our everyday lives and the evidence is all around us. CNC equipment specifically, will be more accessible and known as it continues to shape the world in which we live. There is a lot of growth to come from CNC machines and they will continue to be a major part of the manufacturing industry as a whole.

Chapter 8

CNC Business Related Tips

CNC Business Related Tips

CNC Machinery provides business owners with a variety of benefits, including the opportunity to increase production and lower operational costs. However, if you are new to the world of CNC machinery, integrating them into your manufacturing business can pose a bit of a challenge. Therefore, we've included a short list of tips that help those who are just establishing their CNC manufacturing business, as well as those who are wishing to expand their already existing business.

- Remember that safety is your most valuable asset

Making certain that safety is a priority in your machine shop is imperative. Hiring machine technicians who are skilled and highly trained can help to reduce injuries on the factory floor. Stressing the need for proper machine operation is key, as well as regular maintenance and inspection of the machines.

Though this has been mentioned a number of times throughout the book, maintenance and inspection of your CNC machines is crucial, and can prevent any clogs and malfunctions that can pose possible hazards. Although safety on the factory floor or in the machine shop can result in increased profits and increased productivity rates, maintaining strict safety standards is a must for any manufacturing business owner.

- Don't expand too quickly

If you are just starting or have an already existing machining business, one of the most important things that you can do to build a thriving business is to not expand too quickly. It can be tempting to purchase multiple CNC machines in the beginning. You have to determine whether that is really cost efficient. Also, make sure you have the manpower or budget to expand your facilities to accommodate these new CNC machines.

Instead of quickly expanding, it's better to focus on slow but steady growth that ensures the future success of your business. Even if you have a smaller shop with just a handful of employees, you may still find that you are able to meet your productivity and productivity goals. Therefore, despite the fact that you may believe that purchasing several new machines to begin our expand your machinery business, think twice about whether or not that is going to actually prevent future expansions or decrease revenue growth in the long run.

A good alternative to purchasing brand new CNC machines when beginning CNC machinery or growing your existing manufacturing business is to purchase used CNC machinery. Often, you will find that pre-owned CNC machines can be just as practical, and will give you the ability to carry out a great number of manufacturing processes without the expense of buying a line of new CNC machines.

- Form business partnerships

Forming relationships and partnerships that can help you to start up your machining business or to grow your current business is crucial. Doing so will give you the opportunity to build a profitable networking and support system. Whether you have a friend that may be able to refer clients to your new business endeavor, or another manufacturing shop owner who is willing to offer valuable advice, it's always wise to build business relationships that can increase your chances of succeeding.

One fantastic resource that you have at your disposal, which many other CNC manufacturing shop owners did not have in the past, is the internet. There are a variety of websites and online message boards that are devoted to aiding machinists and machining business owners such as the cncbusinesdirectory.com, surplusrecord.com and many others. Aside from the business relationships and the World Wide Web, you can also choose to ask any family or friends if they might know of anyone who could help you through the process of launching or expanding your business by either sending business your way or by providing you with some helpful insight.

- Focus on your niche in the market

After you have decided which particular parts you are going to manufacture, it's important that you begin to focus on your particular segment of the market. This entails only purchasing machines that you will need for the manufacturing process, marketing to clients within your particular industry as well as offering your services to companies that will purchase your products at a rate that corresponds to your production rate and your turnover. For instance, if you have chosen to manufacture certain automotive replacement parts, then you will want to try to gain clients who will purchase your parts in a quantity that you can actually produce at your shop.

Targeting your particular niche in the market will allow for you to allocate your marketing budget to obtain clients that are more likely to become loyal customers, and to ensure that you are making adequate use of your facilities and your employees. Therefore, choosing a specialty and devoting your marketing and production efforts to becoming a leading manufacturer in that field is a goal that will greatly increase your chances of building a successful manufacturing business.

- Be receptive to evolving technologies

This tip is actually two-fold, as technology can actually help you to build a successful manufacturing company by aiding in your marketing and production goals. Firstly, evolving and emerging technologies can enable you to improve your production processes by increasing the quality of your manufactured items and decrease manufacturing times. Though purchasing newer CNC machinery that includes updated technology can be costly, it can actually be more cost efficient in the long run. This is due to the fact that newer technology can give the ability to make your production processes simpler or carry out new manufacturing tasks that were not possible before.

Technology can also give you the opportunity to team up with tech-savvy marketers who can reach your target market via the internet, or you can even take on the internet marketing yourself. Thanks to social media sites and well designed business web sites, you now have the chance to increase your client base, even if you have a limited marketing budget. Technology can help your manufacturing business in a variety of ways, so it's always a good idea to be receptive to new technologies and to look for any new innovations that can make your manufacturing shop more productive and profitable.

- Look for related products to manufacture

Though it's practical to specialize, as was previously mentioned in this list of helpful tips, this doesn't necessarily mean that you should limit your production if you can easily produce parts using the machines that you currently have. For example, if you notice that a number of your clients are asking for items which you have the potential to make by utilizing the machines that you already have or by simply switching its programs and processes to manufacture these new items, then you may want to think about offering them to clients who are in need of them. Along these same lines, if you have discovered that you may be able to make a sizable profit and increase the success of your manufacturing business by making a part that is associated with the items that you currently manufacture, then you may want to consider purchasing another CNC machine.

- Be aware of your competitors

Being conscious of what your competitors are offering or what technologies they have recently integrated can give you an edge in your industry. Any fluctuations in the market can cause manufacturers to speed up production rates or lower costs in order to keep up with their competition. As a result, you may have to change your production methods or processes in order to stay competitive in your niche market.

- Make machine inspection and maintenance a priority

This is, quite possibly, the most valuable tip that you will ever receive in terms of running a manufacturing business, in that ensuring that your CNC machines are in proper working order and carrying out regular maintenance, increases productivity and potential profits. For a minimal inspection and maintenance cost, you can possibly prevent expensive repairs and extensive machine down-time in the future.

- Invest in quality used machinery whenever possible

If you can purchase quality, pre-owned CNC machinery when starting your CNC manufacturing business or expanding upon your existing business, it's highly recommended. There are a number of dealers (many of which are listed in the directory within this very book) that sell high-end, high-performance, used machinery at a fraction of its original price.

If necessary, use additional CNC equipment to improve the production process. If you find that you can simplify your production process or improve the quality of the parts you manufacture by purchasing additional CNC equipment, you may want to consider doing so. Also, it's highly recommended that you carry out as much of the manufacturing process in-house as possible.

Selling Used CNC Machinery

- Maintenance, inspections and repairs are selling factors

Maintaining your CNC machine and regularly inspecting your CNC machine are the two most important factors when selling your machine. A company looking to buy a CNC machine will surely have it inspected prior to purchase. The machine should not be dinged, scratched, broken with oil leaks, etc. To get top dollar for your CNC machine it is imperative that a business take pride in their equipment and tools. Maintenance and additional costs to repair any CNC machine is always considered when making a purchase. It is vital that maintenance, service and repairs are done on the machine prior to listing the machine for sale as well as having a proper inspection.

- Know equipment value

Here are a few of the most common reasons for knowing value or having an appraisal of your CNC equipment: sale of equipment, refinancing shop equipment, litigation, mergers, sale of your business, purchase of a new machine, tax purposes, insurance, and retirement to name a few.

An appraisal can be a great way to give you an idea of value of your equipment. There are a number of instances in which it is necessary to obtain an appraisal on your CNC equipment or even your machine shop, plant or factory.

You will find the short list of appraisals types in the industry below.

- Liquidation Appraisals
- Forced Liquidation Appraisals
- Fair Market Appraisals
- As-Is Value Appraisals
- Replacement Cost of CNC Machine Appraisals

Chapter 9

Mini Guide to Buying Used Machinery

If you are planning on buying pre-owned or "used" CNC machinery instead of new, then this guide can help give you a basic overview as well as recommendations that may be helpful for buying used CNC machinery.

Benefits of Purchasing Used CNC Machinery

Buying a pre-owned CNC machine can give even those with a rather small budget the opportunity to obtain a high quality piece of CNC equipment. As a general rule, used CNC machines can be discounted by around 25% on average, though that number can greatly increase for various reasons. However, if you make sure that you have the machine properly inspected before you make the purchase, you are greatly reducing the likelihood of being surprised by major unexpected repairs and ensure yourself that you will get a long life out of the machine.

Purchasing a used CNC machine can give you the opportunity to start a machining business without spending a fortune on new machines, and can help already existing manufacturing businesses to easily expand. Be sure that you get the machine inspected as well as verification of regular maintenance. There are a number of benefits that CNC machinery can bring a business without making the significant investment that is usually associated with quality machining equipment.

You should always have a CNC machine inspected before you make your purchase. Even if a seller guarantees that the machine is in good working order, it's better to have a qualified CNC inspection expert thoroughly examine the machine and provide you with an inspection report. Having an inspection performed on used CNC machinery can detect and prevent any issues that may lead to any expensive repairs and can give you the added assurance that you are truly getting what you are paying for.

Tips for Buying Used CNC Machinery

Purchasing used CNC machinery can be a great option, especially if you're new to the world of CNC machinery or on a limited budget. However, when making your purchase there are a number of things that you should keep in mind and a few questions that you should ask the seller, in order to ensure that you get a machine that meets your machining needs and best value for your money.

Here are a handful of helpful tips for buying pre-owned CNC machinery:

- Always have an inspection done before you buy used equipment

This is, quite possibly, the most important tip that you should keep in mind when purchasing a used CNC machine. Always having an inspection performed on the CNC equipment before you actually buy it is imperative, as it will give you an accurate idea of the condition of the machine, how much you may have to spend on parts to fix any issues, and ensure that you are getting exactly what you are paying for.

- Hire an expert to inspect

Hiring an expert to come along with you before you purchase the machine is vital and will give you a better indication of the overall condition of the machine to determine whether or not it is a good value. Every CNC machine has many parts and functions that need to be checked. For example, inspecting the ways or guides, the tool changer, the motor, the switches, the chip conveyor, as well as any other important parts, are extremely important just to name a few.

- Know who you're doing business with

If you are purchasing a used CNC machine, it's always a good idea to buy from a seller who already has a good reputation in the industry such as members of the Machinery Dealers National Association (MDNA) and appraisers of the Association of Machinery and Equipment Appraisers (AMEA). Thanks to the internet, you can now easily check online reviews of a particular pre-owned CNC machine seller and compare a variety of different dealers or brokers who offer high quality, high performance CNC machinery.

- See machine under power

It is good to know if the machine is under power prior to purchase, which will give you a chance to inspect it. The machine might still be in production. It's recommended that you see the machine in operation and if possible run all movable functions at a variety of speeds.

- Have someone there that can test all of the controls

Always make sure they have their operator there to operate the machine. If possible bring someone along with you who can actually operate the machine, so that you are able to determine if it is in proper working order. If you are unable to find someone who can operate the CNC machine, check ahead to see if the seller can arrange for an operator to be on-site during the inspection. Doing so will allow for you to test the machine's functions and parts, such as the live tooling, spindle operation, etc.

- Check the axes in motion

It's wise to check all of the axes of the machine and make sure that they transition smoothly and oscillate well when stopped. Bad ball screws or way covers might produce an irregular noise when the axes are moving. Also, make sure that the servo motors do not make a high pitched noise while in operation.

- Listen for any strange or loud noises while operating

While the operator is controlling the machine during your inspection, listen for any strange noises. Bad bearings will cause a "growling" sound when the machine is running, and a variety of other issues may present irregular noises to alert you that something is possibly wrong with the CNC machine.

- Know hours and usage

There are many other factors that will help determine overall condition such as dings, rust and sounds. You should check the hours on the control or meter, but keep in mind that certain control hours may have been reset. If no hours are available then find other way to determine the amount of usage. Ask whether the machine was running one full eight hour shift a day or three shifts a day. Thoroughly examine the machines parts and functions inside and out.

- Check serial on faceplate

A vast majority of CNC machinery originally comes with a faceplate, which you can find on the back panel of the machine. The serial number of the machine records should correspond with the serial number that is printed on a small faceplate, usually located on the back of the machine. From the serial number the OEM (Original Equipment Manufacturer) can tell you, in most cases, how the machine was equipped when originally purchased, which accessories were originally included with the machine, as well as confirm the original owner.

- Find out the history of the machine

It's highly recommended that you find out as much as possible about the history of the CNC machine you are intending to purchase. Some issues that you may want to clarify with the seller are: the age of the machine, how often it was used, what material the machine cut, how the machine treated as well as where you can find replacement parts or accessories if it's an older machine.

- Ask the seller about any recent maintenance or repairs

You should always ask the seller about any recent repairs or maintenance that has been done on the CNC machine. If so, request the paper work or service history report.

- Ask which type of material they typically used on the machine

Typically, machines that have used tougher materials, such as steel, are going to cause more wear and tear on the machine than lighter materials like aluminum. Thus, it's usually better to purchase a machine that has dealt with lighter materials.

- Ask for any manuals

If you opt to purchase a "used" CNC machine, you should request that the seller also gives you any manuals that they may have for the machine. Manuals can help you to know details regarding parts, programming, recommended maintenance, and electrical components of the machine.

Chapter 10

How much of a discount is typical with used CNC machinery?

Typically, a CNC machine that is used is around 25% discounted when compared to its new counterpart. Depending upon the age, condition and urgency to sell the CNC machine, the discount could be as high as 50% or even more. Therefore, purchasing a pre-owned CNC machine that has been properly maintained or refurbished, is often well worth the investment, and can be an ideal alternative for those who are just starting their manufacturing business or expanding their existing business.

What is required to operate a CNC machine?

It's highly recommended that anyone who operates a CNC machine is skilled and has received training. Though it may be true that CNC machinery is typically less labor intensive and easier to use than manual machinery, it's still important to have a skilled machinist operating the machine and inputting the programs.

Here are some of the basic requirements that a CNC operator typically must meet:

- Knowledge of computers and CNC related applications.

- The ability to produce parts to meet specifications and work on items with accuracy.

- The ability to monitor the operations on the factory floor and to change to quickly adapting environments.

- Extensive knowledge of safety standards pertaining to CNC machines.

- The ability to meet the physical demands that the position calls for, such as carrying of potentially heavy or large items, lifting, twisting, and bending, among other labor intensive actions that may be required on the factory floor.

Is having a CNC machine cost effective even for smaller businesses?

If you own a business that can increase its productivity or profitability by purchasing a CNC machine, then it is worth the investment. It's a common misconception that having a CNC machine is reserved only for large corporations and factories who have unlimited budgets and manpower. In fact, you can easily purchase a CNC machine today for a fraction of their previous cost, especially if you buy a pre-owned or refurbished one. Given that CNC machinery can provide virtually any manufacturing business with a wide range of benefits, having a CNC machine is cost effective and considered to be a great return on your investment, for virtually any production-based business, regardless of their size.

What should I keep in mind when choosing a used machinery dealer?

They have been in business for at least 1 year and have positive online reviews from previous customers.

Their Better Business Bureau record does not have any complaints and they have a positive rating.

They have a facility where they are able to inspect and/or repair CNC machinery, and can offer some sort of reassurance that the machine has been properly inspected prior to the sale.

Many businesses are members of national organizations that adhere to higher standards. The MDNA and AMEA are only two of many organizations that represent trustworthy businesses. Look into these associations and organizations for expertise and exceptional customer service.

What are some ways that associations help the CNC machine industry?

Associations are there to ensure, to the best of their ability, that ethical business practices are maintained in their industry. The MDNA and AMEA are two of many organizations that are known to have reputable industry representatives. The MDNA is an association that you might want to consider when purchasing used equipment. The MDNA is the Machinery Dealers National Association. This association is rooted on ethics and good business practices in the used equipment industry. Their mission is to further the lawful interests of the industry, promote higher business standards and ethics and conducting activities in a manner to improve the objectives of the industry, as stated on their website. Another perk that comes from doing business with a MDNA member is their machine guarantee. The MDNA guarantee is typically a 30-day repair or return privilege covered by the dealer on certified pre-owned equipment. Another organization such as the MDNA is the Association of Machinery and Equipment Appraisers (AMEA). Their standards of professional appraisal practice, ethical conduct, and market-based experience provides more security when deciding to buy or sell used equipment. There are many other organizations that provide services to ensure that good business practices are being met.

How much does an inspection save?

Just to give you an idea of much money you could potentially save by having an inspection performed and carrying out regular maintenance of your CNC machines, here is a brief list of some approximate prices for important CNC machine parts should you have to replace them down the line due to neglect or failure to inspect before making a used CNC purchase.

New spindle replacement can cost between $6,000 and $25,000.

New ball screws can cost replacement between $6,000 and $13,000.

New way cover replacement can set you back $2,000 to $15,000.

New spindle drive replacement can cost $2,000 to $10,000 to replace.

New axis drive can typically range in price from $3,000 to $8,000.

Bear in mind that these prices may or may not include labor or removal of the old damaged parts, nor do they account for the loss of profit that you will have due to lengthy non-productivity times. Whether buying new or used, many costs can be avoided with an inspection.

How should I select a CNC machinery appraiser?

When it comes time to have your CNC machinery appraised, it's important that you choose a CNC appraiser who is experienced and has extensive knowledge of the current market value of your particular CNC equipment's make and model.

Here are a few questions that you should ask before hiring them to appraise your CNC machine:

- How many years have you been a CNC appraiser and are you accredited by the AMEA, USPAP, etc?

- Can you provide documentation of the appraisal process so that I can verify our appraisal should I be audited or for an insurance loss?

- Can you deliver the appraisal to me by my specified deadline?

- Do you adhere to the Principle of Appraisal Practice and Code of Ethics?

What should I ask when choosing an inspection service?

When you are searching for your CNC inspection specialist, it's important to keep a few key things in mind, so that you can rest assured that you are selecting the ideal inspector for your needs.

Here are some questions you should ask when choosing an inspection service for your CNC equipment:

- Are they experienced in the industry?

One of the key things that you should ask the inspection service that you would like to hire is whether or not they have experience with inspecting the type of CNC machinery that you have. Though many inspection services may state that they have been in the business for a number of years, it's always a good idea to mention your machine's brand and model to find out if they have dealt with your equipment type specifically.

- Can they handle the scale of the job?

In order to gauge whether or not an inspection service can adequately cater to your needs, it's wise to find out if they are experienced in dealing with the quantity of machines that you would like inspected. Also, find out if the inspection service can complete the project within your time frame. Some inspection services prefer to only do the job if there are multiple machines to be evaluated, while others simply cannot take on larger jobs that may require advanced equipment or more than one inspector to complete the inspection by the deadline.

- Do they offer other services related to CNC machinery?

Another great way for choosing an inspection service is to ask whether they offer any other services related to the CNC machinery industry. If they sell used CNC machinery or offer appraisal services, then that is good indication that they have the knowledge and insight of current CNC technologies to provide you with a more accurate and thorough inspection of your equipment.

- What's involved in their inspection process?

You should always ask what is involved in the inspection process that the company offers. A thorough inspection of both the inside and outside of the machine is ideal, as well as one that includes a detailed inspection of the parts and overall condition of the machine. Typically the inspection service will have a list or brief description of what their inspection process entails. However, if they don't, be sure to give them a call and ask them to provide you with their evaluation process beforehand.

Chapter 11

CNC Business Service Directory

The next portion of this chapter is a directory of dependable companies that sell and buy used CNC machinery as well as inspection, appraisal and auction services for individual machines or whole plants. This directory is valuable resource guide for those who wish to sell or buy used CNC machines to build or expand their machining business.

Most of the information provided below can be found directly on their business website. Please contact them directly for any questions or assistance.

DIRECTORY

Companies Selling Used CNC Machinery

Arlington Plastics Machinery (www.arlingtonmachinery.com). Arlington Plastics Machinery sells extruders, injection molders, thermo formers, blow molders, granulators and every other type of plastic machinery.

Automatics & Machinery Co (www.automatics.com). Rely on Automatics & Machinery Co. for expertise and assistance whether selling, buying, appraising or auctioning quality used CNC and fabricating machinery.

Clark Machinery Sales (www.clarkmachinerysales.com). Clark Machinery Sales, LLC specializes in CNC machinery. Clark Machinery Sales buys machinery for stock, facilitates consignments, brokers equipment and more.

Emachinetool.com (www.emachinetool.com). EmachineTool has an extensive list of new and used CNC machinery that is available for sale. EmachineTool offers a wide range of quality machine tool brands at direct-to-business pricing.

Grand Marshall Machinery (www.grandmarshallmachinery.com). Grand Marshall Machinery sells quality pre-owned CNC machinery. Grand Marshall Machinery provides an array of services and can help with nearly all you CNC machine needs.

International Machine (www.internationalmachine.net). Whether your goal is to buy or sell an entire manufacturing plant, or simply to relinquish or acquire an individual machine, International Machine can help. International Machine continues to supply and distribute machines, both domestically and internationally with superior customer service.

JS Peters Machinery Sales (www.jspetersmach.com). JS Peters has a wide variety of CNC machinery for sale. They specialize in several brands and types of CNC machines as well as provide other equipment services such as appraisals and liquidation.

Jordan Craig Machinery (www.jordancraigmachinery.com). Jordan Craig Machinery is a one stop shop for the highest quality fabricating and metalworking equipment. They provide services such as appraisals, financing, shipping, installation, training, software, trade-ins and tooling.

KD Capital Equipment (www.kdcapital.com). KD Capital Equipment purchases single CNC machines or entire machinery facilities. There are a number of additional services that KD Capital Equipment offer, from selling to buying and more.

Lee Stevens Machinery (www.stevensmachinery.com). Lee Stevens Machinery has been around since 1948. From engine lathes to CNC drills and saws, they have a wide variety of CNC equipment for sale at very competitive prices.

Lewis Machinery Sales (www.lewismachinerysales.com). Lewis Machinery Sales (LMS) primarily deals in CNC sheet metal fabricating and stamping equipment. Whether CNC or conventional equipment, you can find their full inventory on their website.

Machinery Network (www.machinerynetwork.com). Machinery Network has been in business for over 20 years, and has an extensive list of their used inventory on their site. They specialize in CNC equipment and provide a wide range of services.

Machinery Resources International (www.machineryresources.com). Machinery Resources International buys and sells used CNC machinery, offers CNC equipment appraisals, and lists auctions and liquidations. They cater to the domestic and international machine industry, providing high-quality and affordable used CNC machinery to manufacturing businesses of all types and sizes

Machinery Systems (www.machsys.com). Machinery Systems is one of the largest stocking dealers in the United States and has been in business for over 25 years. You can count on them if you are looking to acquire or sell most any type of CNC machine.

Machinery Values (www.machineryvalues.com). Machinery Values, with more than 28 years of experience assures you that you are dealing with a professional company that knows the industry. Each salesman is a specialist in a specific type of used machinery, which qualifies him to understand your needs and to offer suggestions and alternatives.

Machnet (www.machnet.com). Machnet, Inc. provides the highest-quality CNC machines at exceptional prices. The Machnet team continues to focus on meeting your needs while maintaining an unparalleled reputation for integrity and honesty.

Manufacturing Solutions (www.manufacturesolutions.com). Manufacturing Solutions is able to assist with a wide range of services such as CNC equipment sales, buying pre-owned CNC machine tools and more. They can assist with equipment appraisals, plant liquidation and most any CNC machine need.

Perfection Industrial Sales (www.perfectionmachinery.com). Perfection Industrial Sales has a rather impressive selection of used CNC machinery. From transfer machines to mold presses, Perfection Industrial Sales specializes in used CNC machinery of all types.

Roskelley Machinery (www.roskelley.com). Roskelley Machinery Corporation provides new and used CNC machinery as well as manual machinery. They provide quality service to all types of manufacturing industries, with over 50 years in the machinery business.

Ryan Machine Company (www.ryanmachineco.com). Ryan Machine Co. Specializes in late Model CNC equipment worldwide. With over 60 years of combined experience in Machine tools, Ryan Machine Co. has earned a reputation for fair and honest transactions.

S&M Machinery Sales (www.usedmachinerysales.com). S&M Machinery has been selling and buying machinery since 1977. They also carry a large inventory from CNC Machines to conventional equipment. They sell a variety of brands and types of CNC machines as well as provide other CNC machine services.

Star Equipment Co (www.starequipment.net). Star Equipment Co. sells new and used surplus industrial equipment, machinery, and machine tools. They specialize in metalworking and fabricating machine tools and offer a wide selection of quality equipment.

Sterling Machinery Exchange (www.sterlingmachinery.com). Sterling Machinery buys, sells and trades used CNC machinery in quantities ranging from single machines to entire manufacturing facilities. They sell quality equipment as well as provide other services to meet your CNC machine needs.

Superb Machinery (www.superbmachinery.com). Superb Machinery provides high quality used CNC machinery which can be found on their website, in addition to offering appraisals, auction services and inspection services. They sell a variety of CNC equipment such as horizontal and vertical machining centers, brake presses, turret punches, turning centers, waterjets, lasers also including portable CMM arms and more.

Tramar Industries (www.tramarindustries.com). Tramar Industries specializes in CNC lathes and machining centers as well as wide range of other machine tools. They provide machine tool services around the world and a great resource for trained machine tool professionals.

Used CNC (www.usedcnc.com/). Used CNC, Inc. sells quality pre-owned CNC machinery via their website. The machines that they sell have been thoroughly inspected and checked for quality, durability, as well as functionality. They specialize in providing clients with used CNC machines that will meet their specific needs and focus on building loyal customer relationships by being CNC machinery consultants.

Used Machinery Sales (www.cnctool.com). Used Machinery Sales, LLC has been in business for over 40 years, and has a full inventory of quality pre-owned CNC equipment. They feature used CNC machine centers, metalworking machinery, fabricating machinery and much more.

Wheeler Machinery Sales (www.wheelermachinery.com). Wheeler Machinery Sales has been dedicated to providing the industry with quality used metal working equipment, dependable service, and competitive prices for generations. Wheeler Machinery Sales is one of California's leaders in the used metal working industry.

Yoder Machinery (www.yodermachinery.com). Yoder Machinery is your source for used machine tools. Centrally located in Toledo, Ohio, with a huge selection of machines for sale. Yoder Machinery has been in business for many years and continues to provide machine tool expertise.

Companies Buying Used CNC Machinery

1ˢᵗ Machinery Auctions (www.1stmachineryauctions.com). They offer outright purchases of used CNC machines, private treaty sales, tender sales, and commission auctions, as well as guaranteed auctions. A guaranteed auction gives you the peace of mind of knowing that you will reduce seller risk by being assured a minimum return for your used CNC machinery.

Arlington Plastics Machinery (www.arlingtonmachinery.com). Arlington Plastics Machinery is continually buying used equipment for stock and for our customers. Arlington Plastics Machinery can make cash offers and pay promptly to help you turn your idle equipment into additional funds for your budget.

Blue Star Machinery (www.bluestarmachinery.com). Blue Star Machinery buys used CNC machinery via certified funds or wire transfers, and will provide you with a quick quote. They have a trade-in service that is also available, if you would like to get a newer machine at a greater discount by selling them your used CNC equipment.

CNC South (www.cncsouth.net). CNC South buys and sells used CNC machinery, and will pay cash for quality pre-owned machinery. As an added bonus, they also offer training and installation of your new machine, so that you can have your equipment up and operating as soon as possible.

JS Peters Machinery Sales (www.jspetersmach.com). JS Peters offers a brokerage service, in which you list your used CNC machinery for sale. Newer model turning centers, vertical and horizontal machining centers and grinders are typically prefer, though they broker a wide variety of other machine tools.

K.D. Capital Equipment (www.kdcapital.com). KD Capital Equipment provides cash for your surplus equipment, whether it is a single machine or an entire manufacturing facility. You can fill out a contact form on their site to receive a free quote or offer for your CNC machines, or to explore consignment options to sell your equipment.

Loeb Equipment (www.loebequipment.com). Loeb Equipment purchases quality pre-owned CNC machinery and has a form that you can fill out to receive a quote or offer for your surplus machinery. They also have a list of wanted machinery on their site that includes pieces of equipment that are being requested by clients.

Machinery Network (www.machinerynetwork.com). Machinery Network, in addition to selling and appraising used CNC machines, also purchases pre-owned equipment. You can get a quote for your surplus machinery on their website.

Machinery Resources International (www.machineryresources.com). Machinery Resources International buys and sells used CNC machinery, offers CNC equipment appraisals, and lists auctions and liquidations. They cater to the domestic and international machinery industry, providing high quality, and affordable used CNC machinery.

Machinery Systems (www.machsys.com). Aside from selling quality used CNC machinery, Machinery Systems will buy your used machinery as well. They will purchase single machines or large quantities of CNC equipment, and even offer cash within 24 hours for certain high quality single CNC machines or small quantities.

Prestige Equipment (www.prestigeequipment.com). Prestige Equipment buys surplus CNC machinery from companies who are downsizing or changing out their machinery. They also hold auctions and provide other services to help get you cash or liquidate your equipment.

S&M Machinery Sales (www.usedmachinerysales.com). S&M Machinery buys used steel processing equipment, metalworking machines, and industrial equipment. CNC turning machines, horizontal and vertical machines, grinders, and presses of all kinds are included in the lengthy list of specific CNC machines that they will purchase from sellers.

Star Equipment Co (www.starequipment.net). Star Equipment Co. is always looking for inventory. They will either buy your machine outright for cash or will list it on their site as a consignment. Their sales team has hundreds of customers actively looking for equipment.

Sterling Machinery Exchange (www.sterlingmachinery.com). Sterling Machinery buys, sells and trades used CNC machinery in quantities ranging from single machines to entire manufacturing facilities. If you are looking to sell your equipment fast and easy, Sterling Machinery Exchange can help.

Superb Machinery (www.superbmachinery.com). Superb Machinery is always looking to purchase pre-owned equipment. Superb Machinery provides quality customer service and an expertise you can rely on. Whether looking to sell one machine or a whole plant Superb Machinery can help.

Inspection Services

Focus CNC (www.focuscnc.com). Focus CNC uses high end technology to inspect CNC machinery for a wide variety of industries. They will inspect single CNC machines or entire factories to detect any problems prior to purchase or to assess the current state of your CNC machinery.

Heritage Global Partners (www.hgpauction.com). Heritage Global Partners offers in-house inspections or will refer your inspection requests to their partners nationwide. They have competitive rates and are experience in inspecting a wide variety of CNC machinery that is utilized in virtually every manufacturing industry.

Lincoln CNC Service (www.lincolncncservice.com). Lincoln CNC Service specializes in inspection, preventative maintenance, and service and repair of CNC machines. They will provide you with a detailed report that includes in depth analysis of the inspection, as well as any issues that may have been found during the evaluation.

Superb Machinery (www.superbmachinery.com). Superb Machinery inspects newer CNC machines, as well as vintage CNC equipment, and provides videos and photos in their inspection reports to provide you with valuable information about your CNC machine. This CNC inspection company also sells quality, pre-owned CNC machines via their site.

Machine Appraisals

American Auctioneers Group (www.aagauction.com). The appraisers at American Auctioneers Group are accredited by the Appraisals Guild of America, and are experienced in providing accurate current market valuations due to their daily interaction with CNC machinery of all brands and types.

Ashman Co (www.ashmancompany.com). The Ashman Company offers appraisal services, auction services and used CNC machinery. They have been in business since 1987, and have sold over 300 million dollars in industrial tool machine sales and have team members that belong to numerous associations, such as the AMEA and the NAA.

GoIndustry DoveBid (www.go-dove.com). GoIndustry DoveBid provides valuation advice for equipment, inventory, accounts receivable and real estate. They have access to an asset sales dataset and unique business asset, and conducts valuations in over 30 sectors and 70 countries worldwide.

Grand Marshall Machinery (www.grandmarshallmachinery.com). Grand Machining Company has been around since 1927, and they have a reputation for being a trusted and reliable machining sales and appraisal company. They have an AMEA accredited appraiser on staff that has experience in valuing an array of CNC machinery.

Harris Machine Tools (www.harrismachinetools.com). Harris Machine Tools has certified appraisers on staff that have been accredited by the AMEA. They provide true market value appraisals for individual CNC machinery or entire manufacturing plants, and also purchase surplus assets of all quantities.

International Appraisals (www.internationalappraisals.com). Offering both a USPAP appraisal process and desktop appraisals for those who do not need on on-site appraisal, International Appraisals offers a variety of appraisal services backed by years of experience in the industry. From lease end negotiations to considerations for buying or selling used CNC machinery, they provide CNC equipment and machining appraisal advice for a number of business subjects.

K.D. Capital Equipment (www.kdcapital.com). KD Capital's team of appraisers is AMEA certified and adhere to the industry standards for CNC machinery valuations for a variety of industries. Their knowledge and experience enables them for provide accurate fair market values, and they are willing to travel all over the nation to appraise manufacturing plants and bulk CNC machinery, as well as for smaller quantities of CNC equipment.

Machinery Network (www.machinerynetwork.com). Aside from selling used CNC machines, they also conduct machine appraisals. They have accredited appraisers on staff that can assess the value of CNC machinery for insurance purposes or for those who wish to resell their equipment and need to know the fair market value.

Machnet (www.machnet.com). Machnet, Inc. offers appraisal as well as many other CNC machine services; all while providing appropriate technical expertise. If you need to know the value of your equipment, they can help.

Premier Equipment (www.premierequipment.com). Premier Equipment is experienced in evaluating machinery to determine its value, as well as assessing potential machines that you wish to acquire to calculate their real worth. They have an inventory of used CNC machinery and provide several CNC industry services.

Prestige Equipment (www.prestigeequipment.com). Prestige Equipment has certified equipment appraisers on staff and is a member of the Association of Equipment and Machine Appraisers. They have years of experience in appraising industrial CNC equipment for both small and large companies.

ReSell CNC (www.resellmfg.com). ReSell CNC has experience in appraising both manual and CNC machinery. They can provide you with a fair market value of your machinery for insurance, tax or sale purposes, or simply to give you a good indication of the value of your assets. With over 75 years of combined experience, they offer accurate appraisals that are backed by an in depth knowledge of machining industry.

Roskelley Machinery (www.roskelley.com). Roskelley Machinery Corporation can assure you of ethical and professional appraisal service. They also have Certified Equipment Appraiser (CEA) on Staff. As seen on their site, their mission is to make sure that all customers are treated fairly with all of the personal attention needed.

Sabito Machinery (www.sabito.com). Sabito Machinery, Inc. is accredited by both the AMEA and the USPAP. They are active in used CNC machinery sales market, so they have in depth knowledge of the true value of CNC machines of all brands and models. They also offer auction services, asset remarketing, asset recovery and asset verification.

Superb Machinery (www.superbmachinery.com). Superb Machinery offer appraisal services, and has extensive experience in accurately providing fair market values for a wide variety of CNC equipment, as well as plant appraisals. You can contact them via their site or directly by calling their toll free number.

Machine Auction Services

1ˢᵗ Machinery Auctions (www.1stmachineryauctions.com). 1ˢᵗ Machinery Auctions offers auction listings and you can search through their catalog by machine type. You can place bids, auto bids and view your previous bidding history on their site. They will also occasionally list feature auctions, where you can purchase lots that have been grouped together.

American Auctioneers Group (www.aagauction.com). American Auctioneers Group has an extensive list of auctions for CNC machinery via web cast and/or on-site. They have a rather large computer mailing list to notify their subscribers of upcoming auctions, so it's a good idea to join their mailing list to get information about any CNC machinery that may be going up for sale in the near future.

Accelerated Mfg Services (www.acceleratedmfgbrokers.com). Accelerated Mfg Services offers real estate and manufacturing plant auctions. They feature a listing of online auctions for small and large quantities of CNC machinery for a number of industries, as well as an online inventory of surplus equipment for direct sale.

Capital Recovery Group (www.crgauction.com). Capital Recovery Group's site features numerous online auctions that include machinery from all over the world, which gives you the opportunity to have a wider selection of quality pre-owned CNC equipment to choose from. They have been involved in the machinery auction industry since 1998 and are very customer service focused, and has become a leader of machinery appraisals and auction sales.

GoIndustry DoveBid (www.go-dove.com). This particular site is known as one of the world's largest online auctioneer for used machinery, has an up-to-date list of auctions being held all over the world. From whole plant auctions to individual sale of CNC machinery, GoIndustry DoveBid allows for you to buy or CNC equipment that is utilized in a variety of manufacturing industries.

Hilco Industrial (www.hilcoind.com). Hilco Industrial has auction listings on their site where you can register, view the lots and even place your bid for the equipment in some cases. You can also view the date and time of upcoming auctions, as well as the physical location of the auction.

Hoff Online Auctions (www.hoffonlineauctions.com). Hoff Online Auctions lists auctions that are up-coming for both individual machine sales and machine shop sales of large quantities of equipment. They also provide auction services for those seeking to sell their used CNC machinery or factory, and offer commission, split and guaranteed auction formats.

Machinery Network (www.machinerynetwork.com). Machinery Network has a list of current auctions as well as upcoming auctions being held throughout the United States, as well as a list of plant liquidations. From metal working to fabricating machinery, they have a wide selection of CNC equipment for numerous industries.

Perfection Industrial Sales (www.perfectionmachinery.com). Perfection Industrial Sales has an updated listing of auctions throughout the United States that feature a wide range of CNC machinery, and even some worldwide auction sales. Their active auction listings are quite extensive, and they also offer liquidation listings as well. You can typically find short sale of assets and surplus auctions on their site, as well as negotiated sales.

ReSell MFG (www.resellcnc.com). ReSell MFG is a pre-owned CNC dealer who also offers live auction postings on their website. The auctions are typically from third parties located throughout the United States, and feature a wide range of CNC equipment for all types of manufacturing uses.

Tauber-Arons (www.tauberaronsinc.com). Tauber-Arons, Inc. manages auctions for a wide variety of industrial businesses such as machine tools, woodworking, plastics, construction, medical, food processing, printing and pharmaceutical and many more.

Thompson Auctioneers (www.thompsonauctioneers.com). Thompson Auctioneers is dedicated to providing professional auction services to the industrial and commercial equipment market.

Used Machinery Sales (www.cnctool.com). This used CNC dealer has auctions, where you can purchase various types of quality pre-owned CNC machines for your shop or factory. They also have a large stock of used machinery that can be purchased outright.

Plant Appraisals

American Auctioneers Group (www.aagauction.com). The appraisers at American Auctioneers Group are accredited by the Appraisals Guild of America, and are experienced in providing accurate current market value due to their daily interaction with CNC machinery of all brands and types. They also handle manufacturing plant liquidation appraisals.

Appraisal Economics (www.appraisaleconomics.com). With a full staff of appraisers, CFAs, ASAs, CPAs and engineers, Appraisal Economics provides up-to-date valuations and assigns a senior manager to every appraisal project that they handle to supervise the process. They value both CNC machinery and real estate, so they are ideally suited to carry out manufacturing plant appraisals.

Cambridge Partners & Associates (www.cambridge-partners.com). Whether you would like an appraisal to receive sufficient financing for CNC machinery or for tax purposes, Cambridge Partners has extensive experience in manufacturing plant valuations.

Capitol Appraisal Group (www.capitolappraisalgroup.com). Capital Appraisal Group has provided valuations for the nation's largest manufacturing companies. They have the valuable insight that is required to valuate manufacturing plants in economically turbulent times, and can provide you with an accurate fair market value appraisal for whole business or individual CNC equipment.

EAC Valuations (www.enterpriseappraisal.com). EAC Valuations has been providing appraisals since 1971, and carry out valuations for a variety of industries all over the world. They are an affiliate member of the American Society of Appraisers. They have appraisal expertise in CNC machinery, intangible assets and other types of equipment.

Hennig & Company (www.henniggroup.com). Hennig & Company provides appraisals for all assets within the manufacturing plant and machine shop. They are accredited by the American Society of Appraisers, and value a wide range of business types, from the fabrication shops to large facilities.

Heritage Global Partners (www.hgpauction.com). Heritage Global Partners valuates machinery and equipment, industrial inventory, real estate, and consumer products inventory, making them a good choice for manufacturing plant appraisals of virtually any industry. They provide appraisals for a variety of purposes, including: estate tax, estate planning, accounting, asset based lending, insurance, leasing, litigation, and mergers.

K.D. Capital Equipment (www.kdcapital.com). KD Capital's team of appraisers is AMEA certified and adhere to the industry standards for CNC machinery valuations for a variety of industries. Their knowledge and experience enables them to provide accurate fair market values, and they are willing to travel all over the nation to appraise manufacturing plants and bulk CNC machinery, as well as for smaller quantities of CNC equipment.

Manufacturing Solutions (www.manufacturesolutions.com). Manufacturing solutions is a full service company that knows the value of CNC equipment. They are able to assist with a wide range of services such as Equipment Appraisals, Plant Liquidations and more.

Norcal Valuation (www.norcalvaluation.com). Norcal Valuation provides machine shop, manufacturing and plant equipment appraisals for California and surrounding states. They service a number of industries that include CNC equipment. They have an accredited senior appraiser on staff and specialize in machinery and equipment appraisals.

Steven Wall Appraisal Services (www.stevenwallappraisal.com). With 27 years of experience in the industry, Steven Wall Appraisals has experience in valuating machinery and manufacturing plants. They are accredited by the Appraisal Institute and are a senior member of the American Society of Appraisers.

The Branford Group (www.thebranfordgroup.com). The Branford Group can provide manufacturing plant appraisals to assess the fair market value of the equipment, the forced liquidation value of the facility and the replacement cost of the CNC machinery. They specialize in auctions and valuations and have knowledge of industry trends as well as current values of CNC machinery.

Plant Auction & Liquidation Services

Accelerated Mfg Services (www.acceleratedmfgbrokers.com). Accelerated Mfg Services offers manufacturing plant auctions. They also broker manufacturing companies and machine shops and strive to match machinery plant buyers and sellers.

Capital Recovery Group (www.crgauction.com). Capital Recovery Group's site features numerous online auctions that include machinery from all over the world, which gives you the opportunity to have a wider selection of quality pre-owned CNC equipment to choose from. They have been involved in the machinery auction industry since 1998 and are very customer service focused, and has become a leader of machinery appraisals and auction sales.

GoIndustry DoveBid (www.go-dove.com). This particular site is known as one of the world's largest online auctioneers for used CNC machinery. From whole plant auctions to individual sale of CNC machinery, GoIndustry DoveBid allows for you to buy or CNC equipment that is utilized in a variety of manufacturing industries.

Heritage Global Partners (www.hgpauction.com). Heritage Global Partners caters to a number of industries. With a lengthy list of online auctions, global auctions and web cast auctions, Heritage Global Partners is a great site to visit for manufacturing plant auctions and liquidations.

Hilco Industrial (www.hilcoind.com). Hilco Industrial offers auctions and liquidation sales for used CNC machinery. You can view a complete list of auctions that are to be held throughout the United States, as well as, Liquidation Negotiated or Private Treaty Sales. Some of their auctions are even aired as web-casts, whereby you can watch the auction live from your computer.

Hoff Online Auctions (www.hoffonlineauctions.com). Hoff Online Auctions lists auctions that are available for both individual machine sales and machine shop sales of large quantities of equipment. They also provide auction services for those seeking to sell their used CNC machinery or factory, and offer commission, split and guaranteed auction formats.

Industrial Recovery Services (www.irsauctions.com). Industrial Recovery Services has an extensive list of current auctions and liquidations of whole facilities for those looking to purchase bulk machinery or entire manufacturing plants or machining shops. They specialize in auctions for leasing repossessions, industrial assets and plant closures.

Loeb Equipment (www.loebequipment.com). Loeb Equipment has a list of manufacturing plant auctions and liquidations on their site, featuring numerous types of CNC machinery and varying quantities. They also have CNC machinery rental and leasing services available, as well as financing for qualifying machines.

Machinery Network (www.machinerynetwork.com). Machinery Network gives you the opportunity to buy or sell CNC machinery in real-time auctions. Machinery Network has a list of current auctions as well as up-coming auctions being held throughout the United States, as well as a list of plant liquidations.

Sterling Machinery Exchange (www.sterlingmachinery.com). Sterling Machinery features auctions and manufacturing plant liquidations on their website, and even offer financing for certain qualified CNC machine purchases. They have over 55 years of experience in the industry and are well known for their known for their focus on customer service.

The Branford Group (www.thebranfordgroup.com). In addition to offering single machine auctions, The Branford Group also lists plant auctions on their website. They list online auctions, on-site auctions, as well as auctions that will be web cast.

Directory and Listing Services

CNC Business Directory (www.cncbusinessdirectory.com). The CNC Business Directory (CBD) is dedicated to businesses in the CNC machine industry. CNCBusinessDirectory.com is a great resource for CNC business contact information, industry information as well as CNC career assistance.

Findamachine.com (www.findamachine.com). Findamachine.com is a valuable resource for all things CNC. They feature auction listings, CNC machinery for direct sales as well as informative CNC articles and much more.

Got Machinery (www.gotmachinery.com). From Listing New and Used Metal Working Machines to auction listings, Got Machinery has a number of resources that can help guide or direct you to several different machines and services.

Locator Services (www.locatoronline.com). Locator Services, Inc. provides an array of quality used machinery and pre-owned Machine listings. From used automatic screw machines to a full range of CNC fabricating, tube and wire equipment, LocatorOnline.com can help.

Machinetools.com (www.machinetools.com). This site acts as a directory for a variety of machining services, ranging from sales of used CNC machinery to appraisals. With a comprehensive directory that you can sort, Machinetools.com is definitely worth visiting, and is ideal for those who are looking for auction events, selling their used CNC machinery, buying CNC machinery and more.

Surplus Record (www.surplusrecord.com). Serving the industry since 1924, Surplus Record is one of the world's leading online marketplaces for used and pre-owned industrial equipment. Surplus Record has provided a machine listing directory for used and pre-owned equipment which is a valuable resource in the CNC industry.

The CNC Insider:

Fundamentals of the Computer Numerical Control Machine Industry

By Raul Lepez

For additional information related to this book, please visit the website.

www.TheCNCInsider.com

Additional Helpful Information

MULTIPLICATION TABLE

1	2	3	4	5	6	7	8	9	10	11	12
2	4	6	8	10	12	14	16	18	20	22	24
3	6	9	12	15	18	21	24	27	30	33	36
4	8	12	16	20	24	28	32	36	40	44	48
5	10	15	20	25	30	35	40	45	50	55	60
6	12	18	24	30	36	42	48	54	60	66	72
7	14	21	28	35	42	49	56	63	70	77	84
8	16	24	32	40	48	56	64	72	80	88	96
9	18	27	36	45	54	63	72	81	90	99	108
10	20	30	40	50	60	70	80	90	100	110	120
11	22	33	44	55	66	77	88	99	110	121	132
12	24	36	48	60	72	84	96	108	120	132	144

Table of Time Measure

60	seconds	=	1	minute
60	minutes	=	1	hour
24	hours	=	1	day
7	days	=	1	week
30	days	=	1	calendar month
12	months	=	1	year
365	days	=	1	common year
366	days	=	1	leap year
100	years	=	1	century

Table of Dry Measure

2	pints (pt.)	=	1	quart (qt.)
8	quarts	=	0.86 peck (pk.)
4	pecks	=	1	bushel (bu.)
1	cord	=	128 cu. ft.

Table of Liquid Measure

4	gills (gl.)	=	1	pint (pt.)
2	pints	=	1	quart (qt.)
4	quarts	=	1	gallon (gal.)
31 1/2	gallons	=	1	barrel (bbl.)
2	barrels	=	1	hogshead (hhd.)

Table of Paper Measure

24	sheets	=	1	quire
20	quires	=	1	ream
10	ream	=	1	bale

Table of Linear Measure

12	inches	=	1	foot
3	feet	=	1	yard
16-1/2 ft. (5-1/2 yds.)	=	1	rod
660	feet	=	1	furlong
320	rods (5280 ft.)	=	1	mile

Miscellaneous Measures

12	units	=	1	dozen
12	doz.	=	1	gross
12	gr.	=	1	great gross
20	units	=	1	score
1	hand	=	4	inches
1	fathom	=	6	feet
1	knot	=	6076 feet
3	knot	=	1	league
1	bu. potatoes	=	60 lbs.
1	barrel flour	=	196 lbs.
1	cu. ft. of water	=	7.48 liquid gals.
		and weights 62.425 lbs

Diameter of circle x 3.1416 = circumference
Radius of circle square x 3.1416 = area
Atmospheric pressure is 14.7 lbs.
per sq. in at sea level.
13-1/2 cu. ft. of air weights 1 lb.

Table of the Cubic Measure

1728	cubic inches	=	1	cubic foot
27	cubic feet	=	1	cubic yard
128	cubic feet	=	1	cord of wood
24 3/4	cubic feet	=	1	perch of stone

NOTE : A cord of wood is a pile 8 feet long, 4 feet wide, and 4 feet high. A perch of stone or brick is 16 1/2 feet long, 1-1/2 feet wide, and 1 foot high.

Table of Avoirdupois Weight

16	drams	=	1	ounce (oz.)
16	ounces	=	1	pound (lb)
100	pounds	=	1	hundred-weight(cwt.)
2000	pounds	=	1	ton (T.)
2240	pounds	=	1	long ton(L.T.)

Table of Troy Weight

24	grains (gr.)	=	1	penny-weight (dwt)
20	penny-weights	=	1	ounce (oz.)
12	ounces	=	1	pound (lb.)

Table of Circular Measure

60	seconds (")	=	1	minute (')
60	minutes	=	1	degree (°)
360	degrees	=	1	circumference

A degree of the earth's surface or a meridian = 69.16 miles at the equator.

Table of Apothecaries' Weight

20	grains (gr.)	=	1	scruple
3	scruples	=	1	dram
8	drams	=	1	ounce
12	ounces	=	1	pound (lb.)

Table of Surface Measures

144	sq. in	=	1	sq.ft.
9	sq. ft.	=	1	sq.yd
30 1/4	sq.yds.	=	1	sq.rod.
160	sq.rods	=	1	acre
640	acres	=	1	sq.mile

An acre measures 208.71 ft. on each side.
A section of land is 1 sq.mile.
A quarter section is 160 acre.
A township is 36 sq. miles

CONVERSION TABLE

METERS	YARDS	INCHES
1.000	1.093	39.37
.914	1.000	36.00

CENTIMETERS	INCHES	FEET
1.00	.394	.0328
2.54	1.000	.0833
30.48	12.000	1.000

KILOMETERS	MILES
1.000	.621
1.609	1.000

GRAMS	OUNCES	POUNDS
1.00	.035	.002
28.35	1.000	.0625
453.59	16.000	1.000
1,000.00	35.274	2.205

KILOGRAMS	OUNCES	POUNDS
1.000	35.274	2.205
.028	1.000	.0625
.454	16.000	1.000

LITERS	PINTS	QUARTS	GAL.
1.000	2.113	1.057	.264
.473	1.000	.5	.125
.946	2.000	1.000	.25
3.785	8.000	4.000	1.000

LENGTH

1 meter (m) = 100 cm	=	1,000 mm
1 millimeter (mm)	=	.001 m
1 centimeter (cm)	=	.01 m
1 decimeter (dm)	=	.1 m
1 decameter (dkm)	=	10 m
1 hectometer (hm)	=	100 m
1 kilometer (km)	=	1,000 m

CAPACITY

1 liter(l) = 100 cl	=	1,000 ml
1 milliliter (ml)	=	.001 l
1 centiliter (cl)	=	.01 l
1 deciliter (dl)	=	.1 l
1 decaliter (dkl)	=	10 l
1 hectoliter (hl)	=	100 l
1 kiloliter (kl)	=	1,000 l

WEIGHT

1 gram (g)= 100 cg	=	1,000 mg
1 milligram (mg)	=	.001 g
1 centigram (cg)	=	.01 g
1 decigram (dg)	=	.1 g
1 decagram (dkg)	=	10 g
1 hectogram (hg)	=	100 g
1 kilogram (kg)	=	1,000 g

Additional Helpful Information

Geometric Formulas

Rectangle

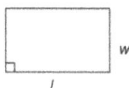

Perimeter: $P = 2l + 2w$
Area: $A = lw$

Square

Perimeter: $P = 4s$
Area: $A = s^2$

Triangle

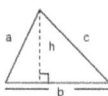

Perimeter: $P = a + b + c$
Area: $A = \frac{1}{2} bh$

Sum of Angles Of Triangle

$A + B + C = 180°$
The sum of the measures of the three angles is $180°$.

Right Triangle

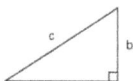

Perimeter: $P = a + b + c$
Area: $A = \frac{1}{2} ab$
One $90°$ (right) angle

Pythagorean Theorem (for right triangles)

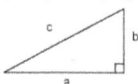

$a^2 + b^2 = c^2$

Isosceles Triangle

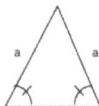

Triangle has two equal sides and two equal angles.

Equilateral Triangle

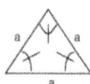

Triangle has three equal sides and three equal angles.

Trapezoid

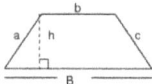

Perimeter: $P = a + b + c + B$
Area: $A = \frac{1}{2} h (B + b)$

Parallelogram

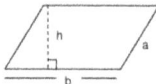

Perimeter: $P = 2a + 2b$
Area: $A = bh$

Circle

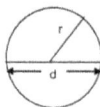

Circumference: $C = \pi d$
$C = 2 \pi r$
Area: $A = \pi r^2$

Rectangular Solid

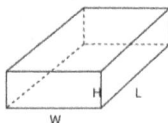

Volume: $V = LWH$
Surface Area: $S = 2LH + 2LW + 2WH$

Cube

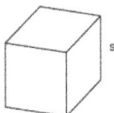

Volume: $V = s^3$

Cone

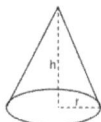

Volume: $V = \frac{1}{3} \pi r^2 h$

Right Circular Cylinder

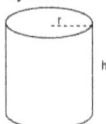

Volume: $V = \pi r^2 h$
Surface Area: $SA = 2 \pi r^2 + 2 \pi r h$

Sphere

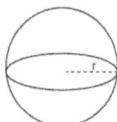

Volume: $V = \frac{4}{3} \pi r^3$

Other Formulas:

Distance: $d = rt$ (r = rate, t = time)
Percent: $p = br$ (p = percentage, b = base, r = rate)

Temperature: $F = \frac{9}{5} C + 32$ $C = \frac{5}{9} (F - 32)$

Simple Interest: $I = Prt$
(P = principal, r = rate, t = time in years)

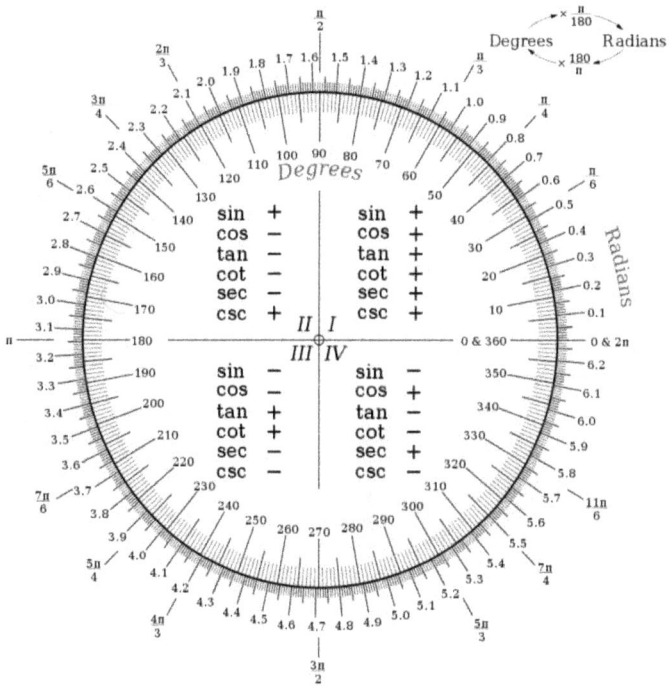

Notes

Notes

Notes